Construction Project Organising

Construction Project Organising

Edited by
Simon Addyman and Hedley Smyth
University College London
UK

WILEY Blackwell

Registered Offices
John Wiley & Sons, Inc., 111 River Street, Hoboken, NJ 07030, USA
John Wiley & Sons Ltd, The Atrium, Southern Gate, Chichester, West Sussex, PO19 8SQ, UK

Editorial Office
The Atrium, Southern Gate, Chichester, West Sussex, PO19 8SQ, UK

For details of our global editorial offices, customer services, and more information about Wiley products visit us at www.wiley.com.

Wiley also publishes its books in a variety of electronic formats and by print-on-demand. Some content that appears in standard print versions of this book may not be available in other formats.

Library of Congress Cataloging-in-Publication Data Applied for

ISBN 9781119807179 [Hardback]

Cover Design: Wiley
Cover Image: © Pauline Lewis/Getty Images

Set in 9.5/12.5pt STIXTwoText by Straive, Chennai, India
Printed and bound by CPI Group (UK) Ltd, Croydon, CR0 4YY

C9781119807179_161222

Contents

Preface

We both first met in 2009 when Hedley was leading the PhD Programme in what was then the Bartlett School of Construction and Project Management, now the Bartlett School of Sustainable Construction (BSSC). Simon was interested in doing a PhD, and we met in the office that Hedley shared with Stephen Pryke to discuss his proposed topic on organisational routines and project organising. In 2014, Simon commenced his part-time PhD under the supervision of Stephen. Over the next five years, we bumped into each other at various points as Simon engaged more in the school both as a PhD student and a lecturer in Stephen's MSc Project and Enterprise Management. In September 2018, Hedley was Simon's internal examiner for his PhD. With great sadness, at the time of Simon's examination, Stephen fell very ill and sadly passed away in early 2020.

In March 2019, Simon formally joined BSSC as Associate Professor of Project Management and took over from Stephen as Programme Director for MSc Project and Enterprise Management, sharing the very office with Hedley where they had first met 10 years earlier. During this time, we exchanged many ideas about research, teaching, and enterprise. Our work is different, both conceptually and from the unit of observation, with Hedley looking mostly at the firm and the perspective of the main contractor and the project interface, and Simon looking mostly at the project and the perspective of the client. Despite these differences, both of our work is influenced by Peter Morris and his Management of Projects paradigm and draws theoretical inspiration from the practice turn in organisational theory.

During the lively and informative discussions we had, Simon explained to Hedley how he and Stephen had written a proposal for an edited book, a plan that was clearly now on hold with Stephen's illness and subsequent passing. Hedley had no hesitation in offering to join Simon in the endeavour, and likewise Simon had no hesitation in accepting this offer. We spent time refining the proposal and engaging in a dialogue about what construction project organising meant to us both and how work in this area had developed over the last 20 years, where it is now, and how it may move forward in the future. We thought not just about theory, but importantly how this applied in practice. What followed from this dialogue has been a very enjoyable journey for both of us and has ended with the publication of this book. We hope your journey with this book is an equally enjoyable one.

We dedicate this book to our dear friend and colleague Stephen.

Simon Addyman and Hedley Smyth, 2022

Editorial – Construction Project Organising: Towards a Theoretical and Practical Understanding

Simon Addyman and Hedley Smyth

1 Introduction

Construction project organising is about process. Organising is active and conceptually distinct from organisation. In practice, organising and organisation cannot be separated. Organising leads to practices becoming embedded. The organisation as an entity, such as a firm or project, once in place in a rudimentary or mature form, frames the scope for any ongoing organising. There is an iterative interaction between the organisation and organising on the ground. It is explicit in this editorial chapter and many of the contributions, and implicit throughout. Organisations do not emerge or arise in a vacuum. While this is self-evident, many contributions from professional bodies of knowledge, textbooks of guidance for those learning and training to be project managers, construction managers, and managers for the firms involved in developing and delivering projects, present the organisation as pre-given – a kind of 'natural' entity that is presented in a taken-for-granted way. Thus, the first major theme of this book is the articulation of *organisation* and *organising*.

Organising is addressed here and in this book as a dynamic process, whether practically or conceptually, through a variety of theoretical lenses. At the crux of organising, it is people who organise themselves and the materials they work with, through saying and doing. This may be undertaken by individuals or by groups and teams. Such organising is undertaken on both an informal, ad hoc basis and on a more formal footing using written rules and procedures to accomplish tasks and reach specific goals. In both the cases, the 'oil' that facilitates people organising is relationships, how they are created and maintained. Multi-organisational teams are formed on temporary bases for and over project life cycles, organisationally acting relatively autonomously from the firms who employ them. Relationships, and the management of them, are therefore central to construction project organising. Both the editors of this book see it following on from and building upon two predecessors where relationships formed a core theme (Pryke and Smyth 2006; Smyth and Pryke 2008). Therefore, a second major theme is extending relationship management through the articulation of *organising and relating*.

There are two important, yet related signals to raise at this point about organising relationship development and organising through relationships. First, there has been considerable emphasis placed on collaboration and collaborative practices across construction that these predecessors reported upon. It is also the case that collaboration has been pursued under overarching contracts and frameworks or merely declared as intentions without it reaching down to the point where the relationships have been managed. Second, much has been left to individual responsibility (Smyth 2015a) and to informal group or team action across the industry and project networks (Pryke 2020).

Organising, as a process that constitutes organisation, creates and recreates structures. Structures can be hierarchical and horizontal. For example, structures for line management and control on the one hand, and structures to facilitate integration at, say, programme management level or through functional departmental boundary spanning on the other hand. These structures both enable and constrain the actions that constitute the organisation. They may be rigid in many cases; yet, they are not static and are subject to change and open to different possible actions (Bourdieu 1990). Subsequent organising acts back to change the structures on the ground. This has been encapsulated sociologically through structuration theory (Giddens 1984), whereby structuring is a continuous process of change and stability, not only as a function of tangible organising on the ground, but also of changes to beliefs and shared values in an organisation within the organisational culture and subcultures.

From a practice-based perspective, a key element of organising is how organisational habits emerge from individuals and groups, which have been called routines (Cohen and Bacdayan 1994). Something that is routine has been conceived as having sufficiently standardised a set of rules that bring order to the processes of organising with minimal management intervention (Cyert and March 1963; Nelson and Winter 1982). More recently, routines have been unpacked, and through enacting the rules, new action patterns may emerge (Feldman and Pentland 2003; Feldman et al. 2021). Emanating from organising, routines are a bridge to structuring the organisation, embedding processes that form structures for the organisation to become recognisable in form. This is not automatic and routines can remain as norms for individuals and small groups without being adopted more widely. Routines, as 'repetitive, recognizable patterns of interdependent actions, carried out by multiple actors' (Feldman and Pentland 2003 : 95), shape the patterning and performing of action (Feldman et al. 2021). They provide the raw material for managing and accomplishing tasks within organisations. In projects and construction, while delivery is temporary in any one location, multiple organisations are involved with different tasks at different life-cycle stages and team membership within these stages repeatedly changes within teams. This means routine dynamics take on special significance for construction project organising as they help inform the dual challenge of adapting to unique situations, while maintaining capabilities (Bygballe et al. 2021; Cacciatori and Prencipe 2021). Thus, a third major theme is the articulation of *organising and routines*.

The significance of the above three themes – *organisation* and *organising*, *organising* and *relating*, and *organising* and *routines* – has played an important role in organising this edited collection, from this introductory editorial chapter to the selection of authors and the content. The chapter proceeds by unpacking the key issues raised in these themes and examining in more detail organisation, project organising, and the extent to which organising may be needed yet absent. We do this in four parts: first, we explore organisation through a layered lens, from the institutional level to the individual. We then explain and position our understanding of project organising and present new ideas and a definition for what we see as *non-organising*. After this, we return to the three themes suggesting avenues for future research for each theme. We finish the chapter with some concluding remarks, which we hope will prompt new discussions in theory and practice.

2 Organisation

2.1 Institutions

The professional bodies, trade associations, client bodies, government policies, and reports trying to stimulate sector transformation have provided some of the main institutional influences on firms and project teams over successive decades. The professional bodies, such as Project Management Institute (PMI), Association for Project Management (APM), and Royal Institute of British Architects (RIBA), have produced plans of work and generated bodies of knowledge. Most of the guidance and associated training is linear in conception, primarily reflecting the tools, techniques, behaviours, and goals of successful completion, especially around time, cost, and quality through a life-cycle model. The idea is largely to bring management certainty to construction projects that are inherently uncertain, spatially unique, and bounded by temporality.

While there is nothing wrong with this as an organising approach, in terms of achieving successful completion, the approach is limited (cf. Morris 2013) as it simply does not embrace the dynamic nature of construction projects in practice (Winter et al. 2006; Blomquist et al. 2010). There have been efforts to adopt more flexible guidance, for example APM (2019) and more recently PMI (2021), both in project management methods, such as agile adopted from information technology (IT) projects (PMI 2017), and to broader management approaches to organisation, including the front end of projects (Morris 2013). However, the dominant approach has remained linear on the basis that certain project management inputs help management create order and reduce uncertainties to induce certain outputs, which are prescriptions that are insufficient in practice and flawed methodologically (Smyth and Morris 2007).

Client and government initiatives have taken a broader perspective, embracing practices around collaborative working, for example, in the UK, the Latham (1994) and Egan (1998) reports, which had considerable international influence in construction. These have had some impact upon improvement at organising as a process. Yet, there are limitations. A limitation noted early on concerned organisational behaviour and the failure of clients and contractors to commit to such practices in coherent and consistent ways (see also Bresnen and Marshall 2000), whereby the initiatives were seen as being as temporary as many large projects and, thus, merely another 'fad' (Green and May 2003; Green 2011).

Another problem is that contractors in particular proclaimed the rhetoric of collaboration, while failing to organise for the active management of relationships (Smyth 2015a; see also Smyth and Edkins 2007). A further criticism of such reports, and indeed all the institutional guidance, is that they are almost solely pitched at the level of the project. The firms that set the organisations that manage the projects in construction have not reformed their management to be transformational in operations since the Second World War in the UK (Smyth 2018) and the pattern is similar elsewhere (Smyth 2022). Similarly, regular clients, such as government or local/statutory organisations, are yet to fully grasp the move from a transactional approach (output focused), to a more value-driven approach (outcome focused) when engaging the market to both develop and deliver projects (see the chapters by Clegg et al., and Xu and Wu).

In sum, the underlying approach to project and construction management tries to set in place organisation along the linear project life cycle that sees organising as a one-off event at the project outset with some acknowledged need for sub-events of organisation (at say, new life-cycle stages, stage gates, or a disruptive event) that largely preserves the idealised and taken-for-granted pre-given organisation. The process of organising at a high level remains largely overlooked and certainly does not occupy a significant and central place in project and construction management. There are positive signs of change in this regard (IPA 2020), lending opportunity for the timeliness of really focusing upon organising here (see also Winch et al. 2022).

2.2 The Construction Firm

Construction firms or companies cover a range of organisational types. First, professional practices can play the lead role as project management organisations providing the service of developing and delivering projects, that is, acting as systems integrator (Prencipe et al. 2003; Davies and Mackenzie 2014). Professional practices or consultant organisations are based upon knowledge workers (Smyth 2011) and are arguably stronger systems integrators than contractors and tend to be more flexible. To that extent, the organisational structures are less rigid, and allowance is built in for organising as a process to accommodate change and coordinate activity, albeit it is the clients that tend to carry the risk for this change.

Tier-1 main contractors remain the predominant construction managers due to their size and available financial resources, in particular working capital and ability to borrow money if and when necessary. That is likely to remain the case for some time. Professional consultants do not have resources and capabilities except among large multidisciplinary international organisations at the top-end. The emergence of IT-based integrated contractor entrants has yet to achieve a fully integrated and mature threat to main contractors or consultants (McKinsey 2017; Smyth 2022).

Therefore, tier-1 main contractors remain dominant for now and hence provide the main focus for this subsection. Many of the points to be raised also apply to many subcontractors with some variance according to specialisation and size. Main contractors are highly transactional, hence reliant upon keeping investment low, using trade credit and where possible interest from working capital between receipts and payment, that is, the return on capital employed rather than return on investment (Gruneberg and Ive 2000; Smyth 2018). The consequence is to organise large firms into divisions around construction subsectors or segments and procurement routes. It also organises functional departments into vertical silos.

This hierarchical organisation provides structural solutions to place staff according to their skills and experience, which helps keep management costs low. When a firm decides to switch to another market, existing divisions may be scrapped or adapted and new staff hired with the requisite skills and experience for the new division. Within departments, costs are kept low by staff being guided by their own expertise with minimum reference to other functions, for example procurement selecting subcontractors without reference to business development as to the detailed profile of value wanted by the client to maximise the outcomes envisaged by the client and end users. Similarly, investment in capabilities, especially programme management capabilities that span projects, is kept to a minimum (Smyth and Wu 2021). This structuring, and the organisation that results, is, when viewed from afar, stable and to an extent rigid. There are rigidities emanating in the silos that reduce internal boundary spanning and integration for the delivery of construction projects (cf. Geraldi 2008).

The primary consequence of this organisation of the firm is that it requires greater organising at the project level. The contractor will provide less explicit guidance or standards as to how to organise each individual project, giving greater autonomy to the project and construction managers to organise each project anew. The service experience that the same clients receive from the same contractor can, therefore, vary extensively. This lack of a structured approach to organising from the firm for each project is taken for granted and even perpetuated by the nature of client procurement approaches. Yet, it considerably adds to the uncertainties on each project and may well add to the cost and time in unaccounted ways. It certainly compromises the service experience and value delivered in the short term and contractor/industry reputation in the long term.

This prime consequence of greater organising being needed at the project level is conceptually located at the firm–project interface.

2.3 The Firm–Project Interface

Any construction project starts within the organisation of the client firm (Winch et al. 2022), at what has been called the front end (Morris 2013). Dealing specifically with client firms (as opposed to private individuals), they cover a significantly heterogeneous group of activities such as housebuilding, commercial development, and infrastructure. These client organisations broadly range from: (i) clients that undertake construction as a one-off event to extend or enhance their business activity (for example, a car manufacturer may build a new production plant), (ii) those that undertake projects as their core business activity (for example, a property developer), (iii) those that do projects alongside their core business activity (for example, an infrastructure owner), and (iv) those that are established for the single purpose of delivering a programme of projects (for example, the UK Olympic Delivery Authority or a special-purpose vehicle). The organisational and management arrangements developing and delivering projects vary hugely across this array of clients from large dedicated departments to single appointed responsible person. This demand side of the equation (Addyman 2020) also includes the wide spectrum of stakeholders, both internal and external, that are an important aspect of construction project organising.

The demand side commences with the client searching for information to structure their problem so that a solution can be found that fits their business model. For repeat clients, this process will likely be led by a sponsoring department, supported by a project management department. They tend to be arranged through portfolios and programmes that interact with corporate governance and management systems to decide on which projects will proceed, when, and with what budget. This process involves engaging a number of construction professionals external to the client firm, such as an architect, and follows either internal procedures or professional guidance to develop the value proposition via a built asset, for example the RIBA Plan of Work (RIBA 2020). Repeat clients, such as major public infrastructure providers, are likely to develop their own internal project management systems and capabilities, to varying degrees of success (NAO 2020; IPA 2019).

In the development stage, the client will engage a number of different professions to develop an agreed design. It is critical in this stage that clients work with suppliers not just to identify the outputs they need, but also the outcomes and benefits and, hence, a clear evaluation through the business case of the value that the project seeks to deliver. The ability of the client to engage the market at this stage is integral to project organising as it creates the early patterns of action from which future capabilities can be built. However, it is not often approached from this capability-building perspective.

The major commitment of the client comes at the transition from development into delivery when major decisions are made in terms of commitment to financing and signing of contracts (Jones and Lichtenstein 2008; Miller and Hobbs 2005). This is done through some form of procurement model that engages the client with the market to select the preferred supplier(s) for design, build, and in some cases maintenance and operation (Hughes et al. 2006). Procurement in the construction industry, therefore, plays a significant role in the overall capability of the project organisation (HM Government 2020). It connects the demand chain with the supply chain (Addyman 2020) and provides a significant part of the structure (rules) around which project teams will organise, both relationally and transactionally.

The capability of the client at the front end to develop a scheme and engage with the market, therefore, becomes a key component; yet, many clients, especially in the public sector, have lost capabilities to do this effectively (Winch and Leiringer 2016) and, hence, are more reliant upon the consultants and contractors. This affects project organising on the demand side and has consequences for whether project organising on the supply side is both adequate or sufficient and appropriate where undertaken (cf. Winch et al. 2022).

On the supply side, organising commences with 'business development', the sales function, whereby a potential or existing client has an identifiable project that is qualified as a prospect to consider bid submission (Smyth 2015b). The task begins with pre-qualifying for the prospect, if the firm is not already on a panel or a long list of potential contractors for the client. The next step is to secure a position on the short list to bid. The process may require some strategic organising and certainly tactical positioning, for example business development managers and other senior management becoming involved in the client qualification process and beginning to assess the content, potential risk profile and mobilising the resources needed to undertake the project, including the (potential) team. In other words, a strategic and tactical assessment is made that forms part of project portfolio management (see the chapter by Killen et al.; Smyth and Wu 2021). As with client firms, the organising around the front end shapes the project both conceptually and practically and effects the way in which any project is subsequently managed. This continues during bid management at a more detailed level, including further negotiations after being awarded the project in principle.

For consultants adopting the project management role, there is typically a smoother transition through these early stages, and organising in the consultancy is typically well integrated as the project gets underway. Among contractors, the process is varied, conceptually affecting integration for delivery. Because of the way contracting firms are structured, the low levels of investment in systems for integration between the firm and project, organising the project during delivery depends upon when the project or construction manager (and their immediate management team colleagues) becomes involved. As well as the way that clients engage the market, there are two factors typically at play. First is the policy or firm strategy as to whether involvement is in the latter stages of business development, during bid management or post-award. Second is the tactical matter of resource availability. Most contractors do not employ the policy of many management consultants that put project managers 'on the bench' between projects to subsequently maximise the fit or alignment with the client and the project; they tend to see who is available at the time to maximise the efficiency of resource utilisation. This decision is often only made, or firmed up, once the contractor knows they have been successful in the bid. This also frequently means that in many cases, construction teams are only fully identified during the subsequent execution stage on site. The consequence is that the later the project or construction manager is involved, the greater the project organising need is likely to be once the project commences, with a consequential risk that any organising and personnel do not align with the project requirements established at the front end. Hence, subsequent organising may be necessary to improve alignment. This will be exacerbated further by key team member changes over the project life cycle, invoking degrees of reorganising, especially where there is a lack of adequate handover. A similar pattern is replicated along supply chains among subcontractors.

Furthermore, consultants and management fee contractors may facilitate greater integration in their own organisations; however, the tendency for ease of management on such contracts is to let large packages to other contractors and key subcontractors, which are beyond the scope of their expertise, automatically triggering a series of subcontracts. In all projects, the tendency has been to grow subcontracting among main contractors to perceptually reduce financial risk for the firm and reduce project risk around content uncertainties (cf. Green 2011). Each fragmentation in delivery prompts degrees of organising at different stages and poses challenges to systems integration and capabilities for effective delivery from the client and user viewpoints. The position is further exacerbated by the use of contract and agency staff among subcontractors, who will to a large extent rely upon self-organising for each task.

At the firm level of main contractors, portfolio management, therefore, rapidly moves to project portfolio management, replicated among subcontractors. The perspective for organising is primarily driven by the needs of each firm rather than the delivery of an integrated service that spans project boundaries. Moreover, contractors in particular tend to follow client programmes rather than manage their projects as programmes. Resources are not invested in the capabilities to develop and improve their own programme management. While client management has seen some improvement and health and safety has also seen considerable improvement, some at programme level, investment remains scant, as the weak implementation or absence of effective knowledge management testifies (e.g. Duryan and Smyth 2019).

Integration and consistency across projects and at the firm–project interface in both clients and contractors is lacking. This has the consequence that project and construction managers act with relative autonomy from the firm, requiring greater project-level organising and self-organising to address problems being encountered (Pryke et al. 2018; see Almadhoob chapter).

2.4 The Project

When a project starts (at the design stage or on site), the content and operational service begins to be delivered. In response to the rules set out in contracts, governance and management plans, project teams tend to act with relative autonomy from their employing firm for some of the reasons provided above. The organising, therefore, will usually start either from scratch or using partially established project routines. The interpretation of the rules by project actors creates processes from which new routines are generated. Except when interfacing with the firm, for example with commercial managers and directors in clients and main contracting organisations, these procedures and routines may be more established. The project and construction managers also have a number of new organisations to take account in their project organising, each with their own culture and systems and, as significant or more significantly, their own respective organising. Conceptually, the organising should account for the management of the inputs and realising effective project outcomes.

From the perspective of a main contractor, there is their own employer organisation to the extent that standard systems and procedures are established, and prior routines established to serve them, these tending to be minimal. Second is the design team and organising the interface with their organisations. Third is the client and satisfying their requirements to deliver valuable outcomes in service experience received and post-completion in use. Fourth is the multiple subcontractor organisations, which are also organising their own subcontractors, contract, and agency operatives.

Across these arrangements and mechanisms, there is considerable organisational complexity and variability of organisational culture, systems, and routines and, importantly, an observable amount of self-organising (Pryke et al. 2018). A great deal of organising could, therefore, be said to be informally instigated, and this is accurate in the context of the autonomy given to the management of each project. However, it would be inaccurate to claim that any emanating routines are 'informal', as the inherent purpose of a routine is to introduce some order and standardisation of action (through the creation of artefacts, such as management plans and contracts), hence immediately blurring on the ground any abstract distinction between the informal and formal.

The relative autonomy of any team varies upon the team and the degree of direct control mechanisms in the form of systems and procedures imposed by the firm. This relationship between the firm and the project changes as the project transitions through the life cycle (Jacobsson et al. 2013). This concept depicts a range of actions, whereby project and construction managers have considerable degrees of scope to heavily influence or determine their organisation. It is not fully controlled vertically from the top down. On some projects, these managers may even take liberty to ignore requirements from their employer. For example, across IT projects, it was found that many project managers did not employ the prescribed project management methodology, for example a prescribed waterfall or agile methodology. The reason provided by project managers was that they had to tailor their methods to client drivers or requirements. On closer questioning a pattern emerged, it repeatedly emerged that the client requirements were not driving the application of the management methods and nor was the project content. The 'hidden' reason was that the methods were managed according to the project manager's own 'comfort zone' (Wells and Smyth 2011). This type of autonomy had some rationale in the sense that the project managers were employing tried and tested routines used previously on other projects. While it was convenient to take forward routines and methods for rapid organising on the forthcoming project, it could lead to misalignments with the rules and needs of the new project and induce inconsistent service across projects from the firm viewpoint (Smyth 2015b).

Fabianski (2017) examined project organisational culture across a range of different urban rail megaprojects in contrasting locations. She found the organising established different types of project organisations and cultures. What was significant is that they changed over the project life cycle and without exception ended up in a hierarchical form. This is important in terms of the systems, procedures, and especially the underpinning routines, which are important cultural artefacts; yet, little is understood of the dynamics of how and why this patterning is established, and indeed, whether it is a norm across all projects. Further, much of the associated management appears to be implicit or reactive in relation to the active management of projects; hence, the benefits and dysfunctions of such shifts in organising are largely unknown. For scholars studying project organising, this becomes a methodological and not just theoretical issue (Leiringer and Zhang 2021).

2.5 Team Members

Construction projects are managed by temporary multi-organisations (Cherns and Bryant 1984). A project team is made up of several project teams not only in terms of each having a different employer, but also having involvement in different disciplines at different stages of the project life cycle. Teams or key team personnel can also change over a project life cycle as new projects are identified or awarded. The power balance between the teams is far from equal because of contracts and durations of involvement. Indeed, subcontractors claim that the cultural, and especially the behavioural and procedural requirements, often imposed by main contractors and sometimes clients add to costs as actions and routines must change to accommodate a variety of project contexts (Duryan and Smyth 2019). On occasions, this is even the case for the same contractor where the construction or project manager has established particular processes on their project.

While the potential organisational dysfunction has been discussed, such changes can also be justified, where the management emphasis upon the skill sets changes over the life cycle. This can be seen through analysis tools, such as Belbin (Belbin 2010). There is a need to explore any link between such changes and changes to the organisation culture where hierarchy often appears to be the preferred organisational form towards the end of the project life cycle. The reason could be, for example, that both the more creative skills to initiate and shape project organising on the ground (plant and shaper categories and the resource investigation skills of Belbin) become less as design and specification are frozen and uncertainties are removed. The implementer and completer finisher categories may work better in hierarchical modes in the latter stages. However, a limitation of such models is obtaining and integrating ideal individual skill sets at different stages. They are heavily reliant on the availability of these skills at the times they are needed. Thus, due to availability constraints, any project team organising will likely involve some form of dysfunctionality that needs managing.

In a drive for more collaborative behaviours, co-location of project teams is one means employed to improve alignment and can both be enhanced by formal collaboration procedures as well as leading to improved collaboration. However, there is also a 'dark side' to co-location that is only beginning to be explored (see the chapter by Aaltonen and Turkuläinen) and is, thus, not a panacea to effective team working.

2.6 Users and Use Value

The evaluation above implicitly shows that organising and the resultant organisation commences with a view to delivering value in terms of the inputs in the design and specification, and importantly through the way in which this is configured by the main contractor in response to the clients' tender document. Systematic integration from the main contractor, client, and across the other multiple organisations (designers, subcontractors, and material and plant suppliers) should also lead to these inputs translating into a valuable completed project for the client and end users. The client may be an intermediary, such as a property developer or investor where the asset value and rental income are key outcomes, or they may be the asset owner, where customer use and satisfaction are key outcomes. Whichever may be the case, all facilities are used by users who have business objectives, policy goals, or social requirements to satisfy.

These outcomes post-completion, the value-in-use, have historically been given less attention. The user and use value of the completed project becomes increasingly overlooked, where the business problem or policy issue of many projects may not be optimally or even sufficiently addressed. Valuable outcomes are not the prime motives around project organising (e.g. Fuentes et al. 2019). At the front end, inputs become more dominant and are employed as surrogates for the outcomes. During the organising of the delivery of the inputs in design and on site, the emphasis increasingly becomes inward-looking, with a focus on fixed times, costs, quality, and meeting the contractually agreed scope (Atkinson 1999) at the fore of decision-making. The emphasis towards a management of projects perspective has influenced front-end activities (Morris 2013), and this plays itself out in practical guidance (IPA 2020). Yet, a consistent focus on value as a key driver for project organising remains at large.

3 Project Organising

The construction project as an organisation is one that can take many different forms. They are generally characterised in the literature as temporary, inter-organisational arrangements and structured with predefined boundaries within which planned action takes place (Lundin and Söderholm 1995; Jones and Lichtenstein 2008; Sydow and Braun 2018; cf. Smyth and Edkins 2007). They sit within a complex *architecture* of organisational arrangements (Denicol et al. 2021; Eccles 1981) and can be understood through a number of different models of economic logic and theoretical lenses, involving both commercial transactions and actor relationships. For example, Bygballe et al. (2013) look at the construction industry as a whole and present four models of economic logic: transaction cost economics, project-oriented, supply-chain-oriented, and network-oriented. While Smyth and Pryke (2008) present four paradigmatic approaches to managing projects, namely traditional, information processing, functional management, and relationships.

We see similar efforts in institutional guidance for practitioners. For example, the UK government's Infrastructure and Projects Authority (IPA) have published a Project Initiation Routemap, with the Delivery Planning Module (IPA 2022) guiding practitioners to create models for operations, delivery, client, and procurement. These models, they say, are interdependent of each other with actions taken in anyone of them impacting on the other. There is then further guidance for these models from a variety of sources. For procurement as an example, the UK Government Construction Playbook (HM Government 2020) seeks to steer the industry towards more value-driven decision-making, which in turn has implications for the organisational structure and the routines participants employ. Questions remain on the widespread competency of clients to follow such guidance.

From a management of projects perspective (Morris 2013), construction project organisations have generally recognisable action patterns, primarily represented by the life-cycle model that dominates both the academic literature and professional standards (Winter et al. 2006; RIBA 2020; APM 2019). Despite the omnipresence of the life-cycle model, from the perspective of routine dynamics in projects (Cacciatori and Prencipe 2021; Bygballe et al. 2021), there remain challenges in recreating the action patterns necessary for collaborative working to flourish (Bresnen et al. 2005; Bygballe and Swärd 2019) as the project moves through its life cycle (Addyman et al. 2020; Locatelli et al. 2020). This is because however rigorously the model is selected and planned for achieving the desired outcomes, in an effort to enact those plans, there is an ongoing need to search for, exchange, and interpret new information. The dynamic processes involved in this may result in unintended action patterns emerging, leading to potentially different outcomes, not all of which are necessarily desirable (Pentland and Reuter 1994; Feldman 2000).

The normative behaviour of the industry has predominantly been on designing and redesigning the ideal models that have a perceived fit with accomplishing the project outcomes (e.g., procurement models; see the chapter by Addyman). From this, teams develop ideal plans (such as activity networks with critical paths) for their sequential and interdependent actions. Yet, the temporary, interdependent, and uncertain nature of construction project organising highlights a very dynamic context that is always in the process of change (Higgins and Jessop 1965; Pryke 2017; Pryke and Smyth 2006). From this viewpoint, and fitting with routine dynamics, whichever model is chosen and entered into, the shared meaning needed for its successful completion is always dependent on the interdependent actions of professional roles, and so open to the tensions arising from coordination in a temporary setting (Bechky 2006; Walker 2015). As we have described above, there is a lack of organising at the firm–project interface leading to the relative autonomy of project and construction managers. Mitigating the problems arising from this autonomous approach is frequently manifested on the ground through the call for multi-organisational collaboration on construction projects.

The construction industry in many countries has long espoused the concept of collaboration as a solution to many reported ills, not least low levels of productivity, without grasping the full potential of its meaning in practice. To address this problem and fitting with current discourse in organisational theory towards a more process and practice perspective (Tsoukas and Chia 2002; Scott and Davis 2017; Feldman 2010), we need to shift our focus '*from*' these static models and '*towards*' furthering our knowledge of the dynamic processes involved in project organising.

This requires a move from thinking about 'organisation' to thinking about 'organising' and the interaction between them. This is a move towards a process as opposed to a substance view of organising, with the former recognising that the models of organisation that we design do not *determine* the outcomes, but create an infinite number of possible action patterns to achieve desired outcomes (Pentland et al. 2020). Process thinking draws greater attention to the dynamic nature of construction projects by exploring the patterning and performing needed to be capable of accomplishing the project (Langley and Tsoukas 2017; Feldman et al. 2021).

4 Non-Organising

The implications of projects and the approaches to the management of projects have been scoped above in relation to organising. The lack of investment from firms (clients and contractors) to support effective project organising has been included, particularly in regard to where this leads at the project level with responsibility for organising on a project-by-project basis. How organising does and can unfold has been scoped. What has yet to be addressed is where there is an absence of organising or, as we term it here, non-organising. Attention is now given to this aspect. Non-organising as a concept poses an academic challenge and needs to be addressed.

Traditional research proceeds on the basis of what is and thus can be observed and recorded. The problem with non-organising, as with non-decision-making for example, is its absence (cf. Bachrach and Baratz 1970). It cannot be seen and directly recorded (Lukes 1974). There may be secondary evidence, for example meeting minutes and reports, but these are likely to be scant, and therefore, indirect evidence may provide a basis. Such evidence is commonly used in natural and social sciences, for example indirect data in areas of quantum physics and psychology.

For non-decision-making, tracing the mobilisation of bias was one means (Bachrach and Baratz 1970), and this phenomenon is seen in project research through its ally, optimism bias (e.g. Flyvbjerg et al. 2003), as well as mobilising other mechanisms of overt and the less obvious so-called hidden controls (e.g. Clegg et al. 2014). To paraphrase Schattschneider (1960, compare with p. 71) in the project context, all forms of project organisation have a bias in favour of the exploitation of some kinds of conflict and the suppression of others, because organising involves mobilising taken-for-granted thinking, bias, and various forms and layers of control and guidance.

Some issues are organised into projects, while others are organised out or simply omitted. Clearly, the actors apply their own values and meanings to the project context and their organising and non-organising, which decision-makers and key actors are able to reflect upon in the context of semi-structured interviews. The researcher similarly has to apply their meaning to interpret the reality of activity derived from any direct and certainly from indirect evidence (e.g. Smyth and Morris 2007). Obtaining such information is nonetheless challenging and is more easily achieved through action research or applying historical methods (e.g. Tuchman 1994) to what is essentially very recent. As Howell and Prevenier state:

> *The historian's basic task is to choose* ***reliable*** *sources, to read them* ***reliably****, and to put them together in ways that provide* ***reliable*** *narratives of the past.* (2001, p. 2, emphasis in original)

A basis can, therefore, be established for recognising and researching non-organising. To begin to scope non-organising for the purposes here, some inferences will be drawn from what we do know and what has already been presented above, which can act as a stimulus for further practical action and research.

At the institutional-level organisations and particularly government and industry, reports tend to focus on the project level, overlooking the important role of the firms as organisations and agents of change through management and investment in capabilities for transformative improvement (e.g. Smyth 2018). Bodies of knowledge also tend to focus on project management as though the project organisation is a pre-given rather than build in the dynamics of project organising.

At the firm level of clients, there remains a predominant trend to engage the market for construction services in a very transactional manner (Oyegoke et al. 2009), which continues to influence organising at the contractor firm level. The use of more relational forms of contracting to achieve collaborative working has improved since the 1990s (Hughes et al. 2006; Mosey 2019; see chapter by Addyman). Additionally, major clients have made efforts to improve their in-house project management capability through portfolio and programme management mechanisms (Winch 2014). These improvements remain underway (e.g. IPA 2020), and yet project front-end capabilities among clients and contractors, and the execution by contractors, remain problematic (NAO 2020; IPA 2019).

Contractor firms keep investment and expenditure at minimal levels to survive in highly transactional high-risk markets driven by bidding regimes. This induces siloed functions internally. More importantly at the project interface, there are low levels of investment to be transformative in terms of project content quality and improving the service experience among clients, contractors, and subcontractors. The consequence of non-organising, or more accurately the lack of effective organising in the firm, is ineffectual systems integration at the project level, hence giving the project team greater responsibility for project organising. Any management rhetoric around collaboration is seldom supported by systems of management, such as behavioural programmes of relationship or interaction management from the firm (e.g. Smyth and Edkins 2007). At the project level, Clegg et al. (see Chapter 1, p. 11) state the following:

> *The assumption of integrated cultures that assumes harmony and coherence is usually an illusion; of far more value is a robust and challenging culture that celebrates its differences and its conflicts and learns from them.*

They go on to state that the individual actors involved typically come from different institutional backgrounds, often nations, and from a variety of organisational types, which tends to introduce differences of viewpoint and conflicts (see also Scott et al. 2011; Jia et al. 2019). The result is a delivery process of power asymmetry between owners and contractors, contractors, and subcontractors, which leads to poor conflict resolution (see also Çeric 2015). Below the cultural level, these are the consequences of poor organising and non-organising at the firm-level and esoteric organising and non-organising at the project level.

Despite the underlying aims to keep investment and expenditure to minimal levels, the outworking of the non-organising and conflicts not only affect quality, but increase subsequent costs to resolve challenges of communication and coordination, and to address conflicts and disputes. Effectiveness and efficiency are compromised, and many of the costs become indirect or unattributable in terms of accountability to finance managers and other senior management in the firms. What is avoided, therefore, is organising for a transformational approach by default and design (see the chapter by Smyth).

At the project level, we scope in more detail what non-organising means practically through the conceptual lens of routine dynamics. Adopting a routine dynamics view is helpful because there are particular practices that can help us understand the tension between the consistency of performance and the need to adapt to the specific project context (Cacciatori and Prencipe 2021; Bygballe et al. 2021). Take, for example, the transition between development and delivery (commitment to execute, securing finance, and signing of contracts); our proposition is that at the outset of this project stage, the specific routines for that specific project are not yet fully established and so the project organisation has only partially established its capability to execute the project. This is because, while people and artefacts bring in routines from the past, for any one particular project, actual patterns of action between the new participants and their new artefacts will take time to become repeatable and recognisable. This repeatability and recognisability is needed for project capability.

The multiplicity of actor perceptions for how to accomplish interdependent tasks and their newness of coming together adds uncertainty to the variety of possibilities for how to proceed (Hærem et al. 2015). This by its very nature involves conflict, overt or covert, large or small, positive or negative. Again, the lens of routine dynamics can help us here as from this perspective, routines act as the mechanism through which dynamic truces can be achieved (Zbaracki and Bergen 2010; Cacciatori 2012; Bygballe and Swärd 2019; D'Adderio and Safavi 2021). Given the relative autonomy of project managers described above, there is a tendency to over rely on the words written in artefacts to give a sense of the ideal meaning on how to proceed to accomplish tasks. Yet, as we have described above, these artefacts, as part of the rules of the project, cannot fully determine the actions of participants.

The relative autonomy at the project level can, therefore, become both a blessing and a curse, a blessing in the sense that competent management teams have the freedom necessary to adapt to new contexts. They can interpret the new 'rules' (contract terms, governance accountability) and create a shared meaning amongst new actors so that effective project organising creates order through dynamic truces. These truces, as routines, emerge through the life of the project and allow actors to achieve its outcomes. A curse is that the new rules to create order are not a guarantee of a truce being achieved in a timely manner.

Prior routines cannot always be relied upon in the new context. Therefore, where actor interpretations of them constrain rather than enable adaptation, shared meaning is not achieved and disorganisation can emerge.

Cunliffe (2003) provides an example here that is relevant to project organising. Citing a discussion with a project planner on the critical path of tasks in a project schedule, Cunliffe notes that sticking rigidly to the idealised schedule and not reflecting on the actualities on the ground led to disorganising. It is explained as follows:

> *Organizing (ordering) aims to subdue disorganizing (dis-ordering), but although we may think we have eliminated disorganization, it will always be present. Thus, organizing is a process of otherness, of dealing with a fundamental tension between organization (presence) and disorganization (absence) (Cooper* 1990*).* (Cunliffe 2003, p. 987)

In this sense, to avoid non-organising leading to disorganisation, the collective intent of actors and their communicative constitution of the project organisation needs managing. It points to the need for project-level organising to be supported by firm-level organising both within the firm itself (client and contractor) and at the project–firm interface. This support needs to take an active role in recreating capabilities through the life of the whole project (Davies and Brady 2016; Davies et al. 2018) and not relying solely on ex ante planned models (and their rules) for idealised action patterns to 'just happen'.

Therefore, we define non-organising as the lack of active management of the project at the interface between the firms and the project. Without this, in an effort to organise collectively, disorganisation emerges, which is not only difficult to recover from but also restricts future performance improvements.

5 Discussion

It is perhaps surprising how little we know about construction project organising, and indeed non-organising. The various bodies of knowledge and industry guidance overlook this dimension. Organising is a dynamic project feature exercising considerable force on the success of project and construction management, not only in terms of time, cost, quality, and scope, but in terms of delivering successful outcomes that satisfy clients and other external stakeholders, as well as internal stakeholder satisfaction that covers financial management and profitability, reputation and brand recognition, learning and job satisfaction, and wellbeing (for example, the chapters by Xu and Wu, and Duryan).

We know something about the triggers that prompt organising and reorganising along construction project life cycles, yet we could know more. We know little as to what individual and shared values prompt key actors to organise or avoid organising. From the little known in project environments, the motives employed are not necessarily following the requirements of the employer, the client, or the project content, but how significant this is and how it unfolds is largely unknown. What shapes the organising is also scantily understood, so in terms of the approaches, styles, and forms of organising, there is a need to unpack what occurs. Thus, the extent of misalignment needs to be researched too and whether particular circumstances, configurations, and triggers are important. The need for this is to begin to more clearly understand what 'good' looks like in terms of construction project organising, which is the vital prior step to knowing how to manage or better manage organising and avoiding dysfunctional non-organising at the firm and project–firm interfaces. The effective and efficient management of organising and avoidance of non-organising would improve performance but can also contribute to a more transformative approach to the management of construction projects.

In the following, we return to the three themes we identified in the introduction to this chapter and take each in turn to discuss future avenues for research into construction project organising. We consider that these avenues are applicable to all the levels we have explored above.

Organisation and organising: research in the domain of construction project organising that we have presented here starts with the onto-epistemological premise that organising constitutes organisation, and the organisation emerges from change (Tsoukas and Chia 2002). Where this change is intentional, organisations exhibit patterns of action, which are an outcome of actors following a set of rules that are either informal and socially tacit in nature or inscribed in the artefacts that actors create and use (Searle 1969; Pentland 1995). These rules provide the boundaries to organising and organisation (Scott and Davis 2017). More empirical work needs to be done on understanding how these rules inform the interaction between actors and their artefacts in constituting the project organisation (Taylor 2011; Dionysiou and Tsoukas 2013).

Organising and relating: We have made clear above that for construction project organising to be successful, relationships and the management of them are a central premise (Smyth and Pryke 2008). We have highlighted some of the pitfalls where there are gaps in relationships and/or relationship management, particularly at the project–firm interface. While further evidence of relationships and their management will be helpful, we suggest that this needs to consider the question '*how do project actors do relating*?' This involves an explicit orientation towards a relational ontology (Emirbayer 1997) and a pragmatist epistemology of inquiry (Lorino 2018). One fruitful avenue of study may be to look at the communicative practices through relating in construction (often held out to be poor) and how they constitute the project organisation (Pryke and Smyth 2006; Söderlund et al. 2008; Schoeneborn et al. 2019).

Organising and routines: We have shown implicitly and explicitly in this chapter how the patterning and performing of actions is foundational for project capabilities. However, organisational routines are little talked about in conventional approaches described in bodies of knowledge and industry guidance. We make an explicit theoretical orientation to routine dynamics, where 'Routine Dynamics involves de-centring and dissolving conventional points of view . . . In a way, decentring is a natural product of looking closely at the patterns of doings and sayings, along with the actors and artifacts that enact them' (Feldman et al. 2021 : 13). The theoretical foundation of and for routine dynamics has been laid out (Feldman et al. 2021), and its application to project organising has been positioned (Cacciatori and Prencipe 2021; Bygballe et al. 2021). What is now required is more evidence from the field on how these routines are reproduced over time and space in construction project organising.

We cannot present these three themes for future research without saying something about methodology. We do not wish to dismiss a multidisciplinary approach to empirical scholarly work; however, the situated and temporal features of construction project organising point to a greater proportion of longitudinal research and a greater emphasis upon ethnographic work. We also need evidence that shows not just that something has changed, but the nature of the change that is taking place by following action sequences and identifying patterns over time and space. This requires an imaginative and creative approach to analysis and writing, whether employing for example ethnography, critical realism or in terms of methods engaged, and action research (cf. Smyth and Morris 2007). Such an approach involves much closer engagement between industry and academia (Jones et al. 2021), sometimes where the researcher and the practitioner are the same (Broft and Addyman forthcoming), where knowledge is mutually developed between the two sides of what is essentially the same coin.

6 Concluding Remarks

Construction project organising is dynamic and needs managing; any stability is only ever temporary. We know that a project organisation needs management; we appoint project managers to do that. Project organising also needs managing, and this cannot achieve its true potential without more work in understanding and developing practices and routines in project-based firms and in projects. But most importantly, a step change in performance is needed at the interface between the project and the firms that create it. This interface is the nexus where firm organising and routines meet project practices and routines and in itself needs routines, which can only be created once the project is underway.

The organisation and management of construction projects is not without its well documented problems in academic, government, and industry publications, some of which are presented in this volume. They generally centre on problems of commercial transactions and problems of human relations. Yet, in producing this book, what is evident in practice is that the industry *says things* and *does things* that: (i) sees these problems as independent of each other and treats them separately; and (ii) seeks panaceas and 'quick fixes' as perceived means of removing these problems. In achieving such objectives, people come to believe they are, and justify their action as being, on the path to transformational change despite past evidence showing time and again this is not the case.

What we have learnt in producing this book is that the industry needs to make a *shift from* redesigning models where these independent problems are perceived to have been removed, *towards* embedding practices and routines at the project–firm interface that enable these problems to be effectively managed to accomplish the desired outcomes for all stakeholders. We believe this involves a rhetorical, conceptual, and practical shift for academics and practitioners.

First, this requires a *rhetorical* shift from *transformation* to *transitioning*. Dictionary definitions of these two words would suggest they have the same meaning. But we would argue that in the construction industry, the rhetorical meaning of the word transformation carries with it a cognitive and performative bias towards the planned and substantial change of something that can be assessed, that is, from steady state A towards steady state B let us say. This has been implicitly or explicitly embedded in bodies of knowledge since the early days of modern project management (Morris 2013). We say state B, because standing back to look at the big picture, industry too often is goal orientated with an implicit idealised steady end-state that is usually unachievable, and therefore, state B is temporary. This is obvious in management but also in projects as evidenced in public–private-partnership-type projects where requirements lead to change in the patterns of use and frequent remodelling too.

Returning to the word transition, it is understood from early anthropological studies as something that is inherently social and, therefore, naturally carries with it ambiguity and uncertainty (Söderlund and Borg 2017). We are not suggesting that something or someone is not transformed, but that a rhetorical shift to transitioning allows us to explore in much more detail the dynamic nature of project organising that we have presented in this chapter. This is made explicit in some of the chapters and is implicitly covered in most other chapters in this volume. We take further guidance here from Lundin and Söderholm (1995) who first positioned transition conceptually in a theory of the temporary organisation.

Understanding before and after states is important. But the meaning of transition as actor perceptions of causal relations (Lundin and Söderholm 1995) points to the internal dynamics of the project and their relation to the more permanent organisations from which projects are created (Jacobsson et al. 2013). *Conceptually,* this is a shift *from* a means–end prospective to an end-in-view perspective. This is a direct conceptual orientation to project organising as an inquiry in the pragmatist sense (Lorino 2018). By this, we mean that managing project organising from the firm and at the project–firm interface starts not with a known or fully defined problem, but with *doubting* that there is a shared understanding of: (i) the collective intent to achieve project outcomes; and (ii) the collective knowledge of how to accomplish interdependent tasks. While we bring in habits, routines, and other practices and structures into the project from previous experiences (some written down in artefacts), a pragmatist inquiry will involve new ways of working, learning, and knowing together as actors take action through the life of the project.

Practically, these new ways of working need managing at the project–firm interface level so as to avoid the pitfalls of non-organising we have described above. Clients and contractors must create firm-level structures and take actions that actively lead to closer integration between the different actors at the project level. These structures must not stay static but continuously adapt to improve performance over future projects. This involves dialogue that shares knowledge, learning, and writing artefacts throughout the project life cycle, a knowing-in-practice approach (Cook and Brown 1999). Blending firm-level structures and routines with project structures and routines is not an easy task and cannot be done independently by any one of the project organisations. The structures and actions must lead to practices that are systematic and collective in nature, which involve understanding how the contract(s) can be collectively interpreted and how the beliefs and values of actors form part of that interpretation, through dialogue.

Project studies (Geraldi and Söderlund 2018), studies about clients (Winch et al. 2022), studies about PBOs (Davies and Hobday 2005), and studies on the construction firm (Smyth 2018) all have good and bad examples of these structures and actions. Most notably, we find good examples where there has been clear intent to collaborate together, transactionally and relationally. But not all these collaborative practices are successful, and we argue they cannot be, if they are not accompanied by the rhetorical and conceptual shift.

Standing back again to look at the big picture, we are saying that changing the structures and systems, indeed restructuring, are insufficient in themselves. Even where the ecosystem or systems are utterly changed, notions that some research seems to be underpinned by, what emerges still needs organising; it needs managing. Others who say it is actually about having the right leadership and leaders, having the right people, are also inadequate for fulfilling this as it will only work if the relationships are effective and if relating induces good organising and routines, using the systems and structures available. Organising construction is inescapable.

References

Addyman, S. (2020). Connecting the 'demand chain' with the 'supply chain': (re)creating organisational routines in life cycle transitions. In: *Successful Construction Supply Chain Management: Concepts and Case Studies* (ed. S. Pryke), 87–108. Chichester: Wiley Blackwell.

Addyman, S., Pryke, S., and Davies, A. (2020). Re-creating organizational routines to transition through the project life cycle: a case study of the reconstruction of London's Bank Underground Station. *Project Management Journal* 51 (5): 522–537.

APM (2019). *The Body of Knowledge*, 7e. Princes Risborough: Association for Project Management.

Atkinson, R. (1999). Project management: cost, time and quality, two best guesses and a phenomenon, it's time to accept other success criteria. *International Journal of Project Management* 17 (6): 337–342.

Bachrach, P. and Baratz, M.S. (1970). *Power and Poverty: Theory and Practice*. Oxford: Oxford University Press.

Bechky, B.A. (2006). Gaffers, gofers, and grips: role-based coordination in temporary organizations. *Organization Science* 17 (1): 3–21.

Belbin, R.M. (2010). *Management Teams*. London: Routledge.

Blomquist, T., Hällgren, M., Nilsson, A., and Söderholm. (2010). Project-as-practice: in search of project management research that matters. *Project Management Journal* 41 (1): 5–16.

Bourdieu, P. (1990). *The Logic of Practice*. Stanford University Press.

Bresnen, M. and Marshall, N. (2000). Partnering in construction: a critical review of issues, problems and dilemmas. *Construction Management and Economics* 18 (2): 229–237.

Bresnen, M., Goussevskaia, A., and Swan, J. (2005). Organizational routines, situated learning and processes of change in project-based organizations. *Project Management Journal* 36 (3): 27–41.

Broft, R., and Addyman, S., (forthcoming). Autoethnography: bridging the project manager-researcher divide. In: Pasian, B., & Turner, J. R. Designs, Methods & Practices for Research of Project Management (2.). *Taylor & Francis*

Bygballe, L.E. and Swärd, A. (2019). Collaborative project delivery models and the role of routines in institutionalizing partnering. *Project Management Journal* 50 (2): 161–176.

Bygballe, L.E., Håkansson, H., and Jahre, M. (2013). A critical discussion of models for conceptualizing the economic logic of construction. *Construction Management and Economics* 31 (2): 104–118.

Bygballe, L.E., Swärd, A., and Vaagaasar, A.L. (2021). A routine dynamics lens on the stability-change dilemma in project-based organizations. *Project Management Journal* 52 (3): 278–286.

Cacciatori, E. (2012). Resolving conflict in problem-solving: systems of artefacts in the development of new routines. *Journal of Management Studies* 49 (8): 1559–1585.

Cacciatori, E. and Prencipe, A. (2021). Project-based temporary organizing and routine dynamics. In: *Cambridge Handbook of Routines Dynamics* (ed. M. Feldman, B. Pentland, L. D'Adderio, et al.). Cambridge, UK: Cambridge University Press.

Cerić, A. (2015). *Trust in Construction Projects*. Routledge.

Cherns, A.B. and Bryant, D.T. (1984). Studying the client's role in construction management. *Construction Management and Economics* 2 (2): 177–184.

Clegg, S., Flyvbjerg, B., and Haugaard, M. (2014). Reflections on phronetic social science: a dialogue between Stewart Clegg, Bent Flyvbjerg and Mark Haugaard. *Journal of Political Power* 7 (2): 1–32.

Cohen, M.D. and Bacdayan, P. (1994). Organizational routines are stored as procedural memory: evidence from a laboratory study. *Organization Science* 5 (4): 554–568.

Cook, S.D. and Brown, J.S. (1999). Bridging epistemologies: the generative dance between organizational knowledge and organizational knowing. *Organization Science* 10 (4): 381–400.

Cooper, R. (1990). Organization/disorganization. In: *The Theory and Philosophy of Organizations* (ed. J. Hassard and D. Pym), 167–197. London: Routledge.

Cunliffe, A.L. (2003). Reflexive inquiry in organizational research: questions and possibilities. *Human Relations* 56 (8): 983–1003.

Cyert, R.M. and March, J.G. (1963). *A Behavioral Theory of the Firm*. Prentice Hall Inc., New Jersey, USA.

D'Adderio, L. and Safavi, M. (2021). Truces and routine dynamics. In: *The Cambridge Handbook of Routine Dynamics* (ed. M. Feldman, B. Pentland, L. D'Adderio, et al.). Cambridge: Cambridge University Press.

Davies, A. and Brady, T. (2016). Explicating the dynamics of project capabilities. *International Journal of Project Management* 34 (2): 314–327.

Davies, A. and Hobday, M. (2005). *The Business of Projects: Managing Innovation in Complex Products and Systems*. Cambridge: Cambridge University Press.

Davies, A. and Mackenzie, I. (2014). Project complexity and systems integration: constructing the London 2012 Olympics and Paralympics games. *International Journal of Project Management* 32 (5): 773–790.

Davies, A., Frederiksen, L., Cacciatori, E., and Hartmann, A. (2018). The long and winding road: routine creation and replication in multi-site organizations. *Research Policy* 47 (8): 1403–1417.

Denicol, J., Davies, A., and Pryke, S. (2021). The organisational architecture of megaprojects. *International Journal of Project Management* 39 (4): 339–350.

Dionysiou, D.D. and Tsoukas, H. (2013). Understanding the (re)creation of routines from within: a symbolic interactionist perspective. *Academy of Management Review* 38 (2): 181–205.

Duryan, M. and Smyth, H.J. (2019). Service design and knowledge management in the construction supply chain for an infrastructure programme. *Built Environment Project and Asset Management* 9 (10): 118–137.

Eccles, R.G. (1981). The quasifirm in the construction industry. *Journal of Economic Behavior & Organization* 2 (4): 335–357.

Egan, J. (1998). *Rethinking Construction: Report of the Construction Task Force on the Scope for Improving the Quality and Efficiency of UK Construction*. London: Department of the Environment, Transport and the Regions.

Emirbayer, M. (1997). Manifesto for a relational sociology. *American Journal of Sociology* 103 (2): 281–317.

Fabianski, C. J. C. (2017). Complex Partnership for the delivery of Urban Rail Infrastructure Project (URIP): How Culture matters for the treatment of Risk and Uncertainty. PhD Thesis. https://discovery.ucl.ac.uk/id/eprint/10040111

Feldman, M.S. (2000). Organizational routines as a source of continuous change. *Organization Science* 11 (6): 611–629.

Feldman, M.S. (2010). Managing the organization of the future. *Public Administration Review* 70: 159–163.

Feldman, M.S. and Pentland, B.T. (2003). Reconceptualizing organizational routines as a source of flexibility and change. *Administrative Science Quarterly* 48 (1): 94–118.

Feldman, M., Pentland, B., D'Adderio, L. et al. (ed.) (2021). *The Cambridge Handbook of Routine Dynamics*. Cambridge: Cambridge University Press.

Flyvbjerg, B., Bruzelius, N., and Rothengatter, W. (2003). *Megaprojects and Risk: An Anatomy of Ambition*. Cambridge University Press.

Fuentes, M., Smyth, H.J., and Davies, A. (2019). Co-creation of value outcomes in projects as service provision: a client perspective. *International Journal of Project Management* 37 (5): 696–715.

Geraldi, J. (2008). The balance between order and chaos in multi-project firms: a conceptual model. *International Journal of Project Management* 26 (4): 348–356.

Geraldi, J. and Söderlund, J. (2018). Project studies: what it is, where it is going. *International Journal of Project Management* 36 (1): 55–70.

Giddens, A. (1984). *The Constitution of Society*. Cambridge, UK: Polity Press.

Green, S.D. (2011). *Making Sense of Construction Improvement*. Chichester: Wiley-Blackwell.

Green, S.D. and May, S. (2003). Re-engineering construction: going against the grain. *Building Research & Information* 31 (2): 97–106.

Gruneberg, S.L. and Ive, G.J. (2000). *The Economics of the Modern Construction Firm*. Basingstoke: Macmillan.

Hærem, T., Pentland, B.T., and Miller, K.D. (2015). Task complexity: extending a core concept. *Academy of Management Review* 40 (3): 446–460.

Higgins, G. and Jessop, N. (1965). *Communication in the Building Industry*. London: Tavistock Institute.

HM Government (2020). *The construction playbook*. https://assets.publishing.service.gov.uk/government/uploads/system/uploads/attachment_data/file/941536/The_Construction_Playbook.pdf

Howell, M. and Prevenier, P. (2001). *From Reliable Sources: An Introduction to Historical Methods*. London: Cornell University Press.

Hughes, W., Hillebrandt, P.M., Greenwood, D., and Kwawu, W. (2006). *Procurement in the Construction Industry: The Impact and Cost of Alternative Market and Supply Processes*. Oxon: Taylor and Francis.

IPA (2019). *Lessons from Transport for the Sponsorship of Major Projects*. London: Infrastructure and Projects Authority https://www.gov.uk/government/publications/lessons-from-transport-for-the-sponsorship-of-major-projects.

IPA (2020). *Improving Infrastructure Delivery: Project Initiation Routemap Handbook*. UK: Infrastructure and Projects Authority http://www.gov.uk/government/organisations/infrastructure-and-projects-authority.

IPA (2022). *Improving Infrastructure Delivery: Project Initiation Routemap – Delivery Planning Module*. London: Infrastructure and Projects Authority https://assets.publishing.service.gov.uk/government/uploads/system/uploads/attachment_data/file/1057652/2641_IPA_Modules_Delivery_Planning_FINAL.pdf.

Jacobsson, M., Burström, T., and Wilson, T.L. (2013). The role of transition in temporary organizations: linking the temporary to the permanent. *International Journal of Managing Projects in Business* 6 (3): 576–586.

Jia, A.Y., Rowlinson, S., Loosemore, M. et al. (2019). Institutional logics of processing safety in production: the case of heat stress management in a megaproject in Australia. *Safety Science* 120: 388–401.

Jones, C. and Lichtenstein, B.B. (2008). Temporary inter-organizational projects: how temporal and social embeddedness enhance coordination and manage uncertainty. Chapter 9. In: *The Oxford Handbook of Inter-Organizational Relations* (ed. S. Cropper, M. Ebers, C. Huxham and P. Smith Ring), 231–255. Oxford: Oxford University Press.

Jones, K., Mosca, L., Whyte, J. et al. (2021). The Role of Industry – University Collaboration in the Transformation of Construction. Transforming Construction Network Plus, Digest Series, No. 4

Langley, A. and Tsoukas, H. (ed.) (2017). *The SAGE Handbook of Process Organization Studies*. Sage.

Latham, M. (1994). *Constructing the Team: Joint Review of Procurement and Contractual Arrangements in the UK Construction Industry*. London: Department of the Environment.

Leiringer, R. and Zhang, S. (2021). Organisational capabilities and project organising research. *International Journal of Project Management* 39 (5): 422–436.

Locatelli, G., Zerjav, V., and Klein, G. (2020). Project transitions – navigating across strategy, delivery, use, and decommissioning. *Project Management Journal* 51 (5): 467–473.

Lorino, P. (2018). *Pragmatism and Organization Studies*. Oxford: Oxford University Press.

Lukes, S. (1974). *Power: A Radical View*. London: Macmillan.

Lundin, R.A. and Söderholm, A. (1995). A theory of the temporary organization. *Scandinavian Journal of Management* 11 (4): 437–455.

McKinsey (2017). *Reinventing construction: a route to higher productivity*. McKinsey Global Institute, https://www.mckinsey.com/business-functions/operations/our-insights/reinventing-construction-through-a-productivity-revolution#.

Miller, R. and Hobbs, J.B. (2005). Governance regimes for large complex projects. *Project Management Journal* 36 (3): 42–51.

Morris, P.W.G. (2013). *Reconstructing Project Management*. Chichester: Wiley.

Mosey, D. (2019). *Collaborative Construction Procurement and Improved Value*. Hoboken, NJ, USA: Wiley-Blackwell.

NAO (2020). *Lessons Learned from Major Programmes*. UK: National Audit Office www.nao.org.uk/report/lessons-learned-from-major-programmes.

Nelson, R.R. and Winter, S.G. (1982). *An Evolutionary Theory of Economic Change*. Cambridge, Mass: Belknap Press.

Oyegoke, A.S., Dickinson, M., Khalfan, M.M. et al. (2009). Construction project procurement routes: an in-depth critique. *International Journal of Managing Projects in Business* 293: 338–354.

Pentland, B.T. (1995). Grammatical models of organizational processes. *Organization Science* 6 (5): 541–556.

Pentland, B.T. and Rueter, H.H. (1994). Organizational routines as grammars of action. *Administrative Science Quarterly* 39 (3): 484–510.

Pentland, B.T., Mahringer, C.A., Dittrich, K. et al. (2020). Process multiplicity and process dynamics: weaving the space of possible paths. *Organization Theory* 1 (3): 1–21.

PMI (2017). *Agile Practice Guide*. Newtown Square: Project Management Institute.

PMI (2021). *PMBoK© Guide*. Newtown Square: Project Management Institute.

Prencipe, A., Davies, D., and Hobday, M. (2003). *The Business of Systems Integration*. Oxford: Oxford University Press.

Pryke, S.D. (2017). *Managing Networks in Project-Based Organisations*. Chichester: Wiley-Blackwell.

Pryke, S.D. (ed.) (2020). *Successful Construction Supply Chain Management: Concepts and Case Studies*. Chichester: Wiley Blackwell.

Pryke, S.D. and Smyth, H.J. (ed.) (2006). *The Management of Complex Projects: A Relationship Approach*. Oxford: Blackwell.

Pryke, S.D., Badi, S., Almadhoob, H. et al. (2018). Self-organizing networks in complex infrastructure projects. *Project Management Journal* 49 (2): 18–41.

RIBA (2020). *Plan of Work*. UK: Royal Institute of British Architects https://www.architecture.com/knowledge-and-resources/resources-landing-page/riba-plan-of-work.

Schattschneider, E.E. (1960). *The Semisovereign People: A Realist's View of Democracy in America*. New York: Holt, Rinehart & Winston.

Schoeneborn, D., Kuhn, T.R., and Kärreman, D. (2019). The communicative constitution of organization, organizing, and organizationality. *Organization Studies* 40 (4): 475–496.

Scott, W.R. and Davis, G.F. (2017). *Organizations and Organizing: Rational, Natural and Open Systems Perspectives*. Abingdon: Routledge.

Scott, W.R., Levitt, R.E., and Orr, R.J. (ed.) (2011). *Global Projects: Institutional and Political Challenges*. Cambridge: Cambridge University Press.

Searle, J.R. (1969). *Speech Acts: An Essay in the Philosophy of Language*, vol. 626. Cambridge University Press.

Smyth, H.J. (ed.) (2011). *Managing the Professional Practice*. Oxford: Wiley-Blackwell.

Smyth, H.J. (2015a). *Relationship Management and the Management of Projects*. Abingdon: Routledge.

Smyth, H.J. (2015b). *Market Management and Project Business Development*. Abingdon: Routledge.

Smyth, H.J. (2018). *Castles in the air? The evolution of British Main contractors*. www.ucl.ac.uk/bartlett/construction/news/2018/feb/castles-air-new-report-charts-management-issues-facing-british-owned-main-contractors (accessed 17 February 2022).

Smyth, H.J. (2022). Transforming the construction firm? In: *Describing Construction* (ed. J. Meikle and R. Best). Abingdon: Routledge.

Smyth, H.J. and Edkins, A. (2007). Relationship management in the management of PFI/PPP projects in the UK. *International Journal of Project Management* 25 (3): 232–240.

Smyth, H.J. and Morris, P.W.G. (2007). An epistemological evaluation of research into projects and their management: methodological issues. *International Journal of Project Management* 25 (4): 423–436.

Smyth, H.J. and Pryke, S.D. (ed.) (2008). *Collaborative Relationships in Construction: Developing Frameworks and Networks*. Oxford: Wiley-Blackwell.

Smyth, H.J. and Wu, Y. (2021) Project portfolio management: practice-based study of strategic portfolio management in the UK construction sector, Bartlett School of Sustainable Construction. www.ucl.ac.uk/bartlett/construction/news/2021/oct/practice-based-study-strategic-project-portfolio-management-uk-construction-sector, University College London.

Söderlund, J. and Borg, E. (2017). Liminality in management and organization studies: process, position and place. *International Journal of Management Reviews* 20 (4): 880–902.

Söderlund, J., Vaagaasar, A.L., and Andersen, E.S. (2008). Relating, reflecting and routinizing: developing project competence in cooperation with others. *International Journal of Project Management* 26 (5): 517–526.

Sydow, J. and Braun, T. (2018). Projects as temporary organizations: an agenda for further theorizing the interorganizational dimension. *International Journal of Project Management* 36 (1): 4–11.

Taylor, J.R. (2011). Organization as an (imbricated) configuring of transactions. *Organization Studies* 32 (9): 1273–1294.

Tsoukas, H. and Chia, R. (2002). On organizational becoming: rethinking organizational change. *Organization Science* 13 (5): 567–582.

Tuchman, G. (1994). Historical social science: methodologies, methods, and meanings. In: *Handbook of Qualitative Research* (ed. N.K. Denzin and Y.S. Lincoln), 306–323. Thousand Oaks: Sage Publications.

Walker, A. (2015). *Project Management in Construction*. Chichester: Wiley.

Wells, H. and Smyth, H.J. (2011). A service-dominant logic – what service? An evaluation of project management methodologies and project management attitudes in IT/IS project business. *Paper presented at EURAM 2011* (1-4 June). Tallinn.

Winch, G.M. (2014). Three domains of project organising. *International Journal of Project Management* 32 (5): 721–731.

Winch, G. and Leiringer, R. (2016). Owner project capabilities for infrastructure development: a review and development of the "strong owner" concept. *International Journal of Project Management* 34 (2): 271–281.

Winch, G., Maytorena, E., and Sergeeva, N. (2022). *Strategic Project Organizing*. Oxford University Press.

Winter, M., Smith, C., Morris, P., and Cicmil, S. (2006). Directions for future research in project management: the main findings of a UK government-funded research network. *International Journal of Project Management* 24 (8): 638–649.

Zbaracki, M.J. and Bergen, M. (2010). When truces collapse: a longitudinal study of price-adjustment routines. *Organization Science* 21 (5): 955–972.

Summary of Chapters

Overview

The aim of this book is to take forward the concept of organising, which not only has considerable applied implications, but also arises from a practice-orientated view of research. Yet, many of the tools and techniques of mainstream project and construction management provide scant help for the research and certainly for practitioners, whether explicitly or implicitly addressing project organising.

We have organised the chapters into four parts that are guided by their overall content, drilling down from the overarching issues to the project level: Part 1 – The Cultural Landscape; Part 2 – Wider Integration; Part 3 – The Firm–Project Interface; and Part 4 – Inside the Project.

Specific chapter aims and objectives are introduced by the invited authors, who in different ways address the interface with organising. The chapters cover a diverse range of topics and conceptual lenses. For each chapter, we briefly summarise the main content in theoretical and empirical terms for both research and practice.

Through the Editorial chapter and the separate contributions, we wish to take forward the understanding of project organising in construction and influence the understanding in generic project environments too. A debate in research and in practice on projects will be a good outcome, yet better still will be agendas towards improvement and transformation. There is no claim here to have picked up every matter of importance in the book, but we are able to generate a list of issues that can inform and be infused into future research.

Part I

The Cultural Landscape

The two chapters in this first part paint a picture of the industry that shows the considerable scope and need for improvement. Importantly, they both offer new ways of working to improve this picture.

In Chapter 1, Clegg et al. take a holistic look at construction culture, finding one that is toxic in nature. Recognising that the shifting temporal and spatial features of construction projects do not fit conceptually with an integrated and stable culture, they present a range of sources of toxicity in the industry.

In Chapter 2, Xu and Wu provide some stark statistics on well-being in the industry and identify similar sources of problems for the lack of a caring approach to the occupational health and well-being of project participants.

Both chapters point to conventional approaches to project research and management being insufficient to meet the needs of the industry and offer potential avenues for how this may change. Clegg et al., for instance, draw on theories of learning and reflection, discussing how a move from the traditional transactional model of construction towards one more closely aligned to an alliancing model can help mitigate this toxicity. Xu and Wu draw on practice theory showing that the same prescriptions to health and safety do not necessarily help the more individual needs of health, safety, and well-being. They equally point to the need for changing the transactional business model towards greater integration, with health, safety, and well-being being actively managed at the project–firm interface through a more bottom-up rather than the conventional top-down approach.

1

Construction Cultures: Sources, Signs, and Solutions of Toxicity

Stewart Clegg[1], Martin Loosemore[2], Derek Walker[3], Alfons van Marrewijk[4] and Shankar Sankaran[2]

[1]*University of Sydney, John Grill Institute and University of Stavanger Business School, Norway*
[2]*University of Technology Sydney, Australia*
[3]*The Royal Melbourne Institute of Technology, Australia*
[4]*Delft University of Technology and BI Norwegian Business School, Oslo*

1.1 Introduction

Early discussions on improving the collective shape of the overall culture of the construction industry revolved around the role that culture plays in construction industry performance and innovation (Anumba et al. 2006). More recent discussions have explored construction culture's impact on the industry's human resource management, corporate social responsibility, and sustainability performance (Azmat 2020). While valuable, cultural change in the sector is slow and challenging, with research being fragmented and in many cases failing to recognise the extent of ambiguity, subcultures, power relations, and the limited boundaries of rational behaviour in the sector. To address this issue, this chapter presents a holistic investigation into construction culture from an organisation studies as well as project management perspective, mobilising the concept of toxic project cultures as a novel conceptual lens to explore new ways to transform the construction industry into a more dynamic, innovative, and socially responsible sector.

1.2 Organisational Culture

The question of what is and is not organisational culture, as well as what its impact is or may be, has been widely debated since the formative work of Schein (1985) and Smircich (1983). Martin (2002) illustrates several approaches in defining an organisation's culture. Each starts from different sets of assumptions, each of which has a significant effect on how culture is subsequently conceptualised. Most frequently, an integrative view of culture is used, which stresses cultural coherence around a unitary frame of reference. Perhaps the

Construction Project Organising, First Edition. Edited by Simon Addyman and Hedley Smyth.

most widely known perspective in this vein is that of Schein, who, in an influential definition, regards organisational culture as:

> *'a pattern of shared assumptions that was learned by a group as it solved its problems of external adaptation and internal integration, that has worked well enough to be considered valid and, therefore, to be taught to new members as the correct way to perceive, think, and feel in relation to those problems.'*
>
> *(Schein 2004, p. 17).*

Schein is joined by many others in this integrative view. For instance, Driskill and Brenton (2011, p. 5) define organisational culture as a socially constructed phenomenon, which defines an organisation's 'unique way of doing things' and the basis for shared meaning and collective action at both conscious and unconscious levels. However, such cultural concordance is highly improbable in practice. While values, assumptions, routines, and meanings may be shared by a (sub)group of people in an organisation, the concept of organisation culture must also include conflict and ambiguity about what is not shared (Martin 2002). It is more appropriate to talk of organisation cultures in the plural or to stress organisations consisting of subcultures, rather than assume a singular cultural consensus (Auch and Smyth 2010; van Marrewijk 2016). Furthermore, organisational cultural is both socially constructed and interactive, continuously being changed by actors in response to an external environment and internal organisational dynamics. When applied to projects, organisation cultures may display unity and coherence in rhetoric, but frequently lack it in practice. Projects are temporary organisations with an acute sense of temporality as well as being composed of a constellation of contested organisational and professional interests and logics with multifarious and plurivocal claims to stakeholding (Jia et al. 2019). Dominant narratives articulating project culture, especially those of elite actors, may superficially stress unity and coherence, while, beneath the surface, project cultures are likely to be influenced by inter-personal and inter- and intra-organisational conflicts between the project owner and client (van Marrewijk et al. 2016) as well as the project management team, contractors, subcontractors, and consultants in supply chains (van Marrewijk 2018).

When combined with intense pressures to perform under tight resource constraints and high levels of uncertainty and risk transfer from clients, such conflicts can nurture cultures of 'fragmentation, antagonism, mistrust, poor communication, short-term mentality, blame culture, causal approaches to recruitment, machismo and sexism' (Ankrah 2007, p. ii), which may be opposed both to legal norms and conventionally accepted codes of behaviour. A portmanteau term for this situation is that of organisational toxicity (van Rooij and Fine 2018). Toxic behaviour condones, neutralises, or enables systematic rule breaking, deviant and damaging behaviours; it disables and obstructs compliance and encourages practices that deviate from expressed values and deflects blame and denies responsibility when these practices are called out (van Rooij and Fine 2018). Toxic practices and behaviours in the construction industry are many. They include, amongst others, corruption, working practices that undermine health and well-being, unfair and opportunistic contractual practices, bullying, intimidation, and discrimination towards various 'out-groups'.

Surprisingly little has been written about the concept of toxic organisational and/or project cultures in the construction industry, despite significant evidence over many decades (Çerić 2016; Lim and Loosemore 2017). Therefore, we next explore how the concept of toxicity can provide new theoretical insights into and explanations of the construction industry's culture, the coping mechanisms that people use to survive it, and strategies that can be used to improve it.

1.3 Toxic Project Culture

Rhetorically, according to Blomquist et al. (2010), managing a project is a performance-based practice 'aiming at the constitution of, coordination and control of [its] activities' (Lundin et al. 2015, p. 3). The traditional engineering-based hard systems view of projects (Blomquist et al. 2010, p. 6) regards these practices as a 'structured, mechanistic, top-down, system-model-based approach relying on systems design, tools, methods and procedures'. In this rational view, the iron triangle of cost, time, and quality is the most basic criterion by which project success is traditionally measured (Pollack et al. 2018), although this is evolving continuously to incorporate softer criteria such as social and environmental value (Loosemore et al. 2021). Blomquist et al. (2010) point out that besides this rational image of how projects are structured, managed, and judged as a success or not, other perspectives are emerging. They point to process- and practice-based perspectives based on the understanding of a project as an accomplishment premised on everyday practices and improvisations. Here, practices are conceptualised as observable actions, meaningfully anchored in organisational cultures (Widler 2001). A practice-based understanding of projects is relevant for conceptualising toxic project cultures as in practice, project delivery constraints, conflicting institutional logics, conflicting subcultures, and poor risk management. Things often go awry, despite project actors being aware and knowledgeable about the value of collaboration and cooperation, including resolving conflicts in a reasonable way. Considerable emotional intelligence, institutional support, and resilient capabilities will be needed to overcome unexpected adversity, features that are frequently missing (Chen et al. 2021). Consequently, project actors too often retreat into self-interest engaging in opportunistic and defensive behaviour, exacerbating fragmentation, further reinforcing the propensity for project culture to take a toxic turn. Project actors respond in diverse ways, ranging from full assimilation into and perpetuation of toxic practices, through coping mechanisms such as humour to open resistance, dissent, and protest through to more informal, illegitimate, and subversive actions (Reed and Loosemore 2012).

1.4 Sources of Toxic Project Culture

There are various potential sources of toxic cultures in construction projects. As noted above, the concept of an integrated stable culture is ill-suited to the circumstances of a construction project in which multiple, temporally shifting participants move in and out of the project as it unfolds within often unrealistic time and budgetary constraints. Typically, this leads to the pursuit of opportunistic behaviours and practices, especially if, within the context of traditional confrontational risk-shifting procurement practices, the project does not have the time or opportunity to develop an integrative culture. In addition, transactional project leadership practices imposing an ends-oriented mentality on the pursuit of project goals can also place intolerable burdens on project members creating toxic cultures exhibiting a range of dysfunctional affective, behavioural, and cognitive coping responses (Zhang et al. 2020). Many aspects of the construction industry's structure, traditions, practices, and norms, including high levels of masculinity, poor risk management, time and cost pressures, including presenteeism as well as bullying, legitimise, normalise, and justify toxic practices, which negatively impact project performance (Galea et al. 2021). Many project managers, despite being excellent in technical skills, simply lack the necessary social skills and emotional intelligence to manage projects, often being unable to regulate their own emotions and temperament (Potter et al. 2018).

Another possible source of toxic culture occurs when the relationship between project owners and project managers, as well as between project managers and contractors, is dysfunctional, resulting in stress for both the parties (van Marrewijk et al. 2016). Stress can be picked up and experienced by other project members and be projected down the line, often leading to the project team experiencing pressure for completion because of project management relational issues. As few projects last a lifetime, project participants experience a reasonable sample of good and bad projects. Consequently, it is reasonable to assume that most project members and stakeholders have a sound idea of how well the project is faring in terms of the toxicity of its management (van Marrewijk et al. 2016).

A final possible source of toxicity is when a construction project is thrown into crisis by unanticipated challenges and events (Loosemore 1999; Blay 2017). Such events highlight hidden inconsistencies and differences in perceptions between actors about where contractual risks lie, destabilising seemingly harmonious project relationships and causing actors to employ a range of legitimate and illegitimate tactics to defend their own interests or even survival. Even in normal times, projects can be characterised by a lack of shared and consistent goals between members and stakeholders (March and Olsen 1976). Projects are, therefore, arenas in which there is a somewhat random confluence of problems, solutions, and participants. Under these circumstances, when unanticipated events occur, actions typically consist of various solutions looking for problems to attach themselves to, to use Cohen et al.'s (1972) characterisation of a garbage can organisation.

Figure 1.1 seeks to conceptualise the potential sources of toxicity, which can become institutionalised in construction projects. It shows that toxicity can be fuelled by external and internal project influences. External influences relate to the business-economic context, whether 'aggressive' targets have been set and are rigorously enforced, particularly if the focus is on 'tight' cost–time budgeting and low bids and a predatory claim-chasing mentality. Professional norms may espouse client focus, but high levels of competition and client value-for-money expectations may undermine good behavioural intentions and interpretation of professional and craft ethical standards (Ruijter et al. 2021).

Project internal influences include structural factors such as how procurement arrangements craft governance and relationship information as well as power asymmetries. Project emotional influences include how people interact, levels of project organisational citizenship behaviour (Ostrom 2015), and affinity with the project (Dainty et al. 2005). Walker and Lloyd-Walker (2020a) explain the influence of skill variety, task identity, and significance

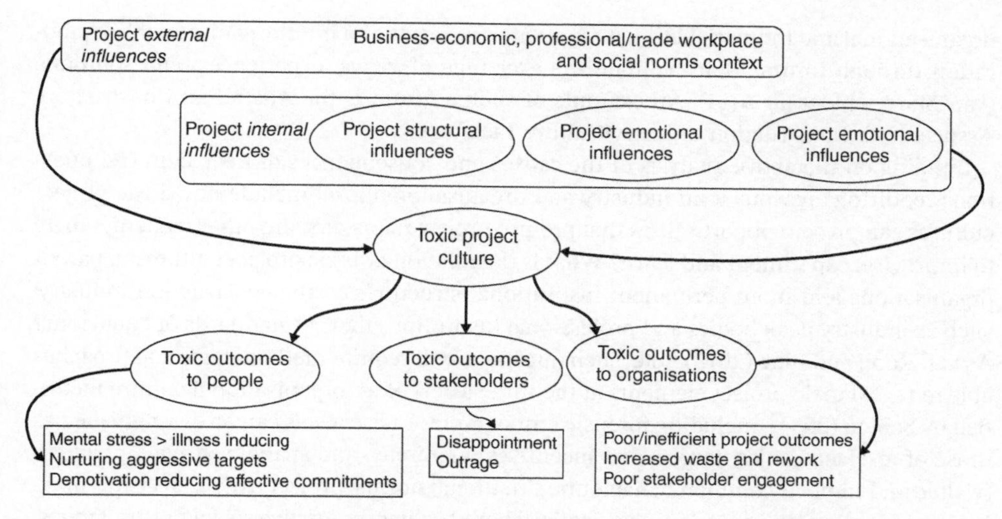

Figure 1.1 Conceptual model of project cultural toxicity and its impact. *Source:* authors.

on how work is interpreted as being meaningful, as well as the extent to which task autonomy is interpreted as work responsibility and how positively task feedback and salience are interpreted. Additionally, levels of physical, mental, and intellectual safety also have an impact on employee well-being, becoming sources of cultural toxicity, influencing project members as well as, potentially, stakeholders, and partner organisations alike (see Figure 1.1). The degree of aggressive compliance to performance expectations demanded by an aggressive project manager or client, as well as how incentives work in the project, is among many routines and other practices influencing how toxically the project is experienced, affecting degrees of member and stakeholder engagement as well as behaviour. Toxicity increases transaction costs for surveillance and control (Haaskjold et al. 2019), accelerating organisational underperformance in general.

1.5 Detoxing a Project Culture

While research points to several strategies that can be employed to reduce problems and impacts, as van Rooij and Fine (2018) found, detoxing an organisation culture requires a fundamental change of underlying structures and processes as well as values and practices enabling and sustaining toxic cultures. An approach promoting transparency, honesty, and responsibility for initiating and sustaining actual cultural change is required (van Marrewijk et al. 2014). Detoxing culture, van Rooij and Fine (2018) argue, begins with an assessment of the full complexity of interactions, structural, processual, and value-based that actors foster, enabling and supporting toxicity. However, in the long term, creating meaningful cultural change will depend not just on addressing these issues but also on ensuring accountability for perpetrators of toxicity, creating safe spaces for victims to contribute to change and forums for future toxicity to be flagged before it has another chance to take root. While doing this at a project level is already challenging (Ruijter et al. 2021), at an organisational and industrial level, it requires cross sector and inter-organisational collaboration through forums that facilitate the exchange of views, experiences, and solutions (van Marrewijk et al. 2014). An example of such a forum is the Australian Constructors Association's Construction Industry Culture Taskforce.[1]

Building on the above analysis of the causes and consequences of toxic cultures, questions requiring answers at an industry and organisational level include how toxic project cultures can present opportunities that prompt project managers and other team members to improvise, experiment and learn? What is the relation of toxic project cultures to parent organisations and more permanent institutional structures in the construction industry, such as industry peak bodies and professional institutions that set standards of behaviour? As van Rooij and Fine (2018) note, such questions will require methodological approaches able to reveal toxic project elements at the three levels of an organisational culture identified by Schein (2004), including forensic ethnography. These levels can be distinguished as those of artefacts (rules, targets and incentives, hierarchy, and spatial settings), practices (visible and implicit common behaviours, situational norms, project rituals, decision-making, and agenda-setting) and values (explicit shared values, injunctive social norms, taboos, and hidden assumptions).

Traditional research tools such as surveys and structured face-to-face interviews are restricted in their ability to reveal these aspects of toxic project culture. They are limited to what we can easily measure, while the very small amount of research insight that we already have defines the questions to be asked. Surveys are not very effective at getting beneath the surface of organisational life into informal institutions and practices that often drive behaviour in toxic cultures. Furthermore, in such a nascent and exploratory field of research, they provide little if any opportunity for researchers to explore unexpected leads into important and unpredictable elements of toxic project cultures. Forensic ethnography requires an interpretivist epistemology that involves the use of in-depth qualitative methods enabling close interaction between respondents and researchers, using qualitative social science research methods and ethnography (Ybema et al. 2009). Despite drawing on deep ethnographic knowledge, detoxing the construction industry will be challenging. The process of assessing a project culture as toxic may undermine successful interventions as it is likely to prompt a backlash from those with vested interests in the status quo. Many project institutions and practices leading to toxic behaviour, whether formal or informal, are likely to be deeply ingrained, sticky, invisible, and resistant to change.

All levels of culture will need to change, and to be effective, attention on the part of project leadership to the change process is required on an everyday basis (Alvesson and Sveningsson 2016). Changing only surface manifestations, those of artefacts and symbols, will not root out problems and eradicate toxic cultures. The underlying processes of inclusion and exclusion, the practices of decision-making and agenda-setting, as well as deeply embedded values frame cultures. If incoming leadership is genuinely different in their values and commitment to cultural change from strongly normalised local project sentiments, they will clash. When trying to hold executives responsible, let alone prosecuting them, one of the unintended consequences of doing so can cause cover ups and blame-shifting and

---

[1] https://www.constructors.com.au/initiatives/construction-industry-culture-taskforce/.

other avoidance behaviours further strengthening the toxic culture. Even when structures and leaders are successfully changed, lower-level managers and employees may obstruct the change implemented, since it is often the case that everyone is involved in rule breaking to some degree (van Marrewijk 2018). Detoxing is likely to require unlearning deeply engrained practices and cognitive restructuring of basic values and assumptions (Schein 2004). Unlearning involves very different cognitive processes to learning and can be especially challenging when it comes to organisational cultural factors, which are deeply embedded (Chandra and Loosemore 2011). A sense of learned helplessness can also develop and spread over time in organisations where power asymmetries make toxic cultures appear inevitable and normalised.

1.6 Stimulating Reflection and Learning

Toxic cultures characterise projects at their worst; what about projects at their best? Projects attuned to reflection and organisation learning are clearly better places in which to work than those projects that foster a toxic culture as well as being more likely to produce better project outcomes. As project cultures coalesce around various issues and events, the diversity of project participants can be turned into a project learning advantage (Senge 1992). The best projects manage their diversity as deliberately directed through purposive strategies. Examples include Olympic projects (Clegg et al. 2002; Pitsis et al. 2003) and the building of the museum of New Zealand Te Pap Tongerwa (Freeder et al. 2021, p. 139), embracing Maori heritage in the design and development of the museum. The people in these projects were driven by the vision of the project, resulting in a project design that enacted a culture premised on learning and deliberation in pursuit of an ultimate and shared value rather than a sectional interest, such as bringing in a project profitably.

In projects, technical errors and failures are *normally* detected and the causes identified to ensure they are not repeated in future (Argyris and Schön 1978). Such single-loop learning takes place within the existing rules, policies, and structures of the project organisation and is essentially corrective. Double-loop learning by contrast requires thinking and learning to do different things to arrive at project goals differently, changing social actions and relations. Doing this is not easy and is constrained by some methods of procurement in the construction industry (Newcombe 1999). Bateson (1958) conceived of another specific type of learning that he called deutero-learning, often referred to as 'triple-loop learning', which involves 'learning to learn' (Schön 1975, p. 8). Projects vested in this mode of learning recognise, reflect on, and change what they are doing by disrupting assumptions, creating discontinuities. Such project learning relies on cultures in which curiosity and innovation are valued and practiced (Ruijter et al. 2021). Projects that prize routines and rational plans as if they were black letter law, rather than as facilitating dynamic practices and capabilities (Parmigiani and Howard-Grenville 2011), will find it much harder to develop emergent approaches, to attend to the flow of process, to be able to improvise as events require. Where a project culture strives hard for top-down integration on the project leaders' terms, such leaders will be less likely alert to opportunities to learn as communication flows will tend to be one-way.

Excellent examples of a learning project culture are integrated project delivery forms such as alliancing and partnerships (Ruijter et al. 2021). Ruijter et al. (2021) highlight radical process innovation in designing and creating a construction project culture to support a public–private partnership through a series of reflection sessions. Walker and Rowlinson (2020) argue that the organisational structure and governance of an alliance contract is designed to overcome many potential toxicity problems. Doing so is not without problems, as Reed and Loosemore (2012) discuss in their discussion of the differences in project culture between alliance and traditionally procured construction projects and the culture shock that construction professionals must navigate when they move from one to another and back again (see Figure 1.2).

Two designed-in alliance contract elements address the toxicity of project culture. First, multi-party forms of agreement integrating design, construction delivery, facility operation, and project owner are necessary for an alliance project culture to create an integrated team for project delivery (Transport 2011). Contract terms stress that an integrated team will execute the work and realise the project output, instead of each contract partner being accountable for their separate part (Ross et al. 2014). Doing so inspires a sense of governmentality that draws the project's teams together (Clegg et al. 2002; Pitsis et al. 2003; Clegg 2019; Ninan et al. 2019). Alliance governance arrangements reduce power and information asymmetry between project participants (Andersen et al. 2020), resulting in project participants sharing a commitment to achieve a best-for-project outcome (Morwood et al. 2008).

Genuine collaboration is premised on alliance participants' respecting each other's expertise and abilities and behaving openly and honestly in their dealings (Morwood et al. 2008). Teams with disparate professional cultures adjust and align their values to a common core, as reported upon in a Finnish alliance project by Matinheikki et al. (2019). Effective and intense collaboration is also facilitated by a no-blame trusting project organisational environment (Lloyd-Walker et al. 2014; Walker and Lloyd-Walker 2020b). No-blame trust is not a pre-given, but has to be developed and maintained throughout the project in a reciprocal relationship between project partners (Ruijter et al. 2021).

All these properties were evident in Pitsis et al.'s (2003) and Clegg et al.'s (2002) study of a major piece of infrastructure associated with the 2000 Olympics in Sydney. The project had

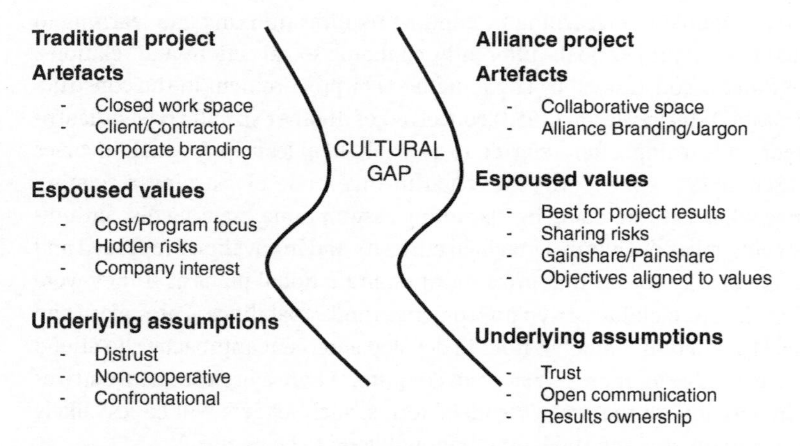

Figure 1.2 Culture gap between alliance and traditionally procured project organisations. *Source:* Reed and Loosemore (2012).

several effective elements constitutive of a more integrated culture. Starting at the outset, the project owner, Sydney Water, chose the alliance partners based on the degree of cultural openness and flexibility they displayed in the tender stage. The tender stage was unconventional; there was no detailed brief, just a statement of purpose to build a storage tunnel on the north side of Sydney Harbour that could capture extreme rain events and prevent the sewage system overflowing, with its contents polluting the harbour and its environs. The tender document was largely composed of photos of the harbour glistening in the sunshine interspersed with shots of the contents of the sewer as well as street detritus in the harbour.

The tender process involved a process in which the degrees of cultural flexibility and openness of the tendering representatives was the main object of scrutiny. The partners were selected based on their willingness to work according to a risk/return contract. Independent auditors had costed what a 'business as usual' project delivery would cost and how long it would take. The latter was important; this project had to be completed in slightly more than half of 'normal time' if it was to be ready for the Olympics. The project owner realised this would take considerable innovation in the project. A risk/return case was based on the independent audit; if the project came in positively on *all* the key performance indicators for the project, there would be a sharing of the profits accrued by saving on the audit's projected costs. These indicators were cost, schedule, occupational health and safety, ecology, and community. Cost and schedule need no explanation. Occupational health and safety should not; in this case, they were particularly important. The construction workers' union was made a stakeholder in the risk/return calculus, so that it would be a beneficiary of innovation in the project. Weekly toolbox meetings were held in which innovative ideas were promoted that could be turned into practice. The harbour is a fragile ecology that the project sought to improve; measures of normal turbidity were a benchmark against which the project, involving barges to ship spoil to a trucking site, was measured. Community was important because the harbour's north side has many affluent suburbs fronting on the water and several measures were taken to ensure that the project did not diminish social capital in these communities.

The project processes that ensued were centred on a specifically designed project culture with 10 points ordered around 'whatever's best for project' autonomously designed by the alliance partners. In the project, each indicator had a regularly rotated champion; for instance, engineers might find themselves championing ecology and ecologists engineering. The organisation learning that ensued was invaluable as it integrated knowledge and sentiment across the disciplinary silos.

The lesson is simple: project organising that is premised on the assumption of an integrated culture that is defined top-down and assumed to be shared because project management has assumed that it is far less likely to produce excellent results than project organising that is premised on creative and constructive conflict, robust 'idea work' (Coldevin et al. 2019) producing deliberated outcomes to which the project adheres precisely because they have been deliberated rather than imposed and assumed. The assumption of integrated cultures that assumes harmony and coherence is usually an illusion; of far more value is a robust and challenging project culture that celebrates its differences and its conflicts and learns from them.

1.7 Conclusions

Alliancing mitigates cultural toxicity (Walker and Lloyd-Walker 2020b). Genuine dialogue between project participants in a low power-information asymmetry environment, in which project participants are expected to challenge assumptions, debate, and explore design and delivery proposals so that they fully understand proposed ideas and can innovate them further (Walker and McCann 2020), creates positivity through robust perspective-taking probing ideas. Senge (1992, p. 241) argues that in effective deliberative dialogue, 'surprising outcomes are achieved that are superior to any one person's or group's starting position'. A key question for future research that addresses project culture that extends beyond alliancing is the extent to which effective deliberative dialogue can be an antidote to toxic culture.

Where projects strive to be highly innovative, they will learn from conflicts, learn from differences, appreciating that any complex project is never *a* culture but a melange of different cultures and subcultures that, rather than being forced into harness and integration with each other, can be sources for deliberations from which pragmatic integration and a degree of tolerance for variance emerges. There may be lessons here that might make more projects more innovative. Thus, there are implications and applications for practice that are twofold. First, our discussion informs practitioners that the cultural transformation of the construction industry is a laborious but necessary process. For example, to contribute to the transition towards a circular economy, construction firms must reflect upon the emergence and acceptance of new practices related to changing organisational roles and responsibilities. Inter-organisational strategic change projects can serve as 'temporary trading zones' (Lenfle and Söderlund 2015), in which actors from different organisations bring in different work practices, narratives, norms, and values, thus creating opportunities for experimenting, knowledge exchange, and changing behaviour. In these arenas, doing things in unusual ways should always be on the agenda, to unlearn ingrained routines. Unlearning involves very different cognitive processes to learning (Chandra and Loosemore 2011). Second, successful as well as unsuccessful projects can function as drivers for change within parent organisations by pressuring shifting, frequently informal, rules within the dominant regime. Project actors bringing in their newly learned practices can spread change within their own organisations and thus contribute to transformation.

References

Alvesson, M. and Sveningsson, S. (2016). *Changing Organizational Culture: Cultural Change Work in Progress*. London: Routledge.

Andersen, B.S., Klakegg, O., and Walker, D.H.T. (2020). IPD governance implications. In: *Routledge Handbook of Integrated Project Delivery* (ed. D.H.T. Walker and S. Rowlinson), 417–438. Abingdon, Oxon: Routledge.

Ankrah, N. (2007). Investigation into the impact of culture on construction project performance. PhD Thesis. University of Wolverhampton.

Anumba, C.E., Dainty, A., Ison, S. et al. (2006). Understanding structural and cultural impediments to ICT system integration: a GIS-based case study. *Engineering, Construction and Architectural Management* 13: 616–633.

Argyris, C. and Schön, D.A. (1978). *Organizational Learning: A Theory of Action Perspective*. Reading, Mass: Addison-Wesley.

Auch, F. and Smyth, H.J. (2010). The cultural heterogeneity of project firms and project teams. *International Journal of Managing Projects in Business* 3 (3): 443–461.

Azmat, Y. (2020). Influence of human factors on the safety performance in Saudi Arabian construction industry. PhD Thesis. University of Salford.

Bateson, G. (1958). *Naven: A Survey of the Problems Suggested by a Composite Picture of the Culture of a New Guinea Tribe Drawn from Three Points of View*. Stanford, CA: Stanford University Press.

Blay, K.B. (2017). Resilience in Projects: Definition, Dimensions, Antecedents and Consequences. PhD Thesis, Loughborough University.

Blomquist, T., Hällgren, M., Nilsson, A. et al. (2010). Project-as-practice: in search of project management research that matters. *Project Management Journal* 41 (1): 5–16.

Çerić, A. (2016). *Trust in Construction Projects*. London: Routledge.

Chandra, V. and Loosemore, M. (2011). Communicating about organizational culture in the briefing process: case study of a hospital project. *Construction Management and Economics* 29 (3): 223–231.

Chen, Y., Chen, Y., Smyth, H. et al. (2021). Enforcement against contract violation in Chinese construction projects: impacts of trust and perceived intentionality. *Construction Management and Economics* 39 (8): 687–703.

Clegg, S.R. (2019). Governmentality. *Project Management Journal* 50 (3): 266–270.

Clegg, S.R., Pitsis, T.S., Rura-Polley, T. et al. (2002). Governmentality matters: designing an alliance culture of inter-organizational collaboration for managing projects. *Organization Studies* 23 (3): 317–337.

Cohen, M.D., March, J.G., and Olsen, J.P. (1972). A garbage can model of organizational choice. *Administrative Science Quarterly* 17 (1): 1–25.

Coldevin, H.G., Carlsen, A., Clegg, S.R. et al. (2019). Organizational creativity as idea work: intertextual placing and legitimating imaginings in media development and oil exploration. *Human Relations* 72 (8): 1369–1397.

Dainty, A.R.J., Bryman, A., Price, A.D.F. et al. (2005). Project affinity: the role of emotional attachment in construction projects. *Construction Management and Economics* 23 (3): 241–244.

Driskill, G.W. and Brenton, A.L. (2011). *Organizational Culture in Action: A Cultural Analysis Workbook*. Thousand Oaks, CA: Sage.

Freeder, D., Sankaran, S., and Clegg, S.R. (2021). The Museum of New Zealand Te Papa Tongarewa: a labour of love and learning. In: *Megaproject Leaders: Reflections on Personal Life Stories* (ed. N. Drouin, S. Sankaran, A. van Marrewijk and R. Müller), 139–149. Cheltenham: Edward Elgar Publishing.

Galea, N., Powell, A., Salignac, F. et al. (2021). When following the rules is bad for wellbeing: the effects of gendered rules in the Australian construction industry. *Work, Employment and Society* 36 (1): 119–138.

Haaskjold, H., Andersen, B., Lædre, O. et al. (2019). Factors affecting transaction costs and collaboration in projects. *International Journal of Managing Projects in Business* 13 (1): 197–230.

Jia, A.Y., Rowlinson, S., Loosemore, M. et al. (2019). Institutional logics of processing safety in production: the case of heat stress management in a megaproject in Australia. *Safety Science* 120: 388–401.

Lenfle, S. and Söderlund, J. (2015). *Projects as temporary trading zones: a theoretical proposal and empirical illustrationIRNOP* (ed. A. Davies), 21–24. London: University College London.

Lim, B.T.H. and Loosemore, M. (2017). The effect of inter-organizational justice perceptions on organizational citizenship behaviors in construction projects. *International Journal of Project Management* 35 (2): 95–106.

Lloyd-Walker, B.M., Mills, A., and Walker, D.H.T. (2014). Enabling construction innovation: the role of a no-blame culture as a collaboration behavioural driver in project alliances. *Construction Management and Economics* 32 (3): 229–245.

Loosemore, M. (1999). A grounded theory of construction crisis management. *Construction Management and Economics.* 17 (1): 9–19.

Loosemore, M., Alkilani, S.Z., and Murphy, R. (2021). The institutional drivers of social procurement implementation in Australian construction projects. *International Journal of Project Management* 39 (7): 750–761.

Lundin, R.A., Arvidsson, N., Brady, T. et al. (2015). *Managing and Working in Project Society: Institutional Challenges of Temporary organizations.* Cambridge: Cambridge University Press.

March, J.G. and Olsen, J.P. (1976). *Ambiguity and Choice in Organizations.* Bergen: Universitetsforlaget.

van Marrewijk, A.H. (2016). Conflicting subcultures in M&A: a longitudinal study of integrating a radical internet firm into a bureaucratic telecoms firm. *British Journal of Management* 27 (2): 338–354.

van Marrewijk, A.H. (2018). Digging for change: change and resistance in interorganizational projects in the utilities sector. *Project Management Journal* 49 (3): 34–45.

van Marrewijk, A., Veenswijk, M., and Clegg, S.R. (2014). Changing collaborative practices through cultural interventions. *Building Research and Information: The International Journal of Research, Development and Demonstration* 42 (3): 330–342.

van Marrewijk, A.H., Ybema, S.B., Smits, K.C.M. et al. (2016). Clash of the Titans: Temporal organizing and collaborative dynamics in the Panama Canal megaproject. *Organization Studies* 37 (12): 1745–1769.

Martin, J. (2002). *Organizational Culture: Mapping the Terrain.* Thousand Oaks, CA: Sage Publications.

Matinheikki, J., Aaltonen, K., and Walker, D.H.T. (2019). Politics, public servants, and profits: institutional complexity and temporary hybridization in a public infrastructure alliance project. *International Journal of Project Management* 37 (2): 298–317.

Morwood, R., Pitcher, I., and Scott, D. (2008). *Alliancing, a Participant's Guide – Real Life Experiences for Constructors, Designers, Facilitators and Clients.* Brisbane: AECOM.

Newcombe, R. (1999). Procurement as a learning process. *Journal of Construction Procurement* 5 (2): 211–220.

Ninan, J., Clegg, S.R., and Mahalingam, A. (2019). Branding and governmentality for infrastructure megaprojects: the role of social media. *International Journal of Project Management* 37 (1): 59–72.

Ostrom, E. (2015). *Governing the Commons: The Evolution of Institutions for Collective Action*. Cambridge: Cambridge University Press.

Parmigiani, A. and Howard-Grenville, J. (2011). Routines revisited: Exploring the capabilities and practice perspectives. *Academy of Management Annals* 5 (1): 413–453.

Pitsis, T.S., Clegg, S.R., Marosszeky, M. et al. (2003). Constructing the Olympic dream: a future perfect strategy of project management. *Organization Science* 14 (5): 574–590.

Pollack, J., Helm, J., and Adler, D. (2018). What is the Iron Triangle, and how has it changed? *International Journal of Managing Projects in Business* 11 (2): 527–547.

Potter, E.M., Egbelakin, T., Phipps, R. et al. (2018). Emotional intelligence and transformational leadership behaviours of construction project managers. *Journal of Financial Management of Property and Construction* 23 (1): 73–89.

Reed, H. and Loosemore, M. (2012). Culture shock of alliance projects. *Association of Researchers in Construction Management (ARCOM): 28th Annual Conference,* Edinburgh, United Kingdom, (3–5 September), 543–553.

van Rooij, B. and Fine, A. (2018). Toxic corporate culture: assessing organizational processes of deviancy. *Administrative Sciences* 8 (3): p23.

Ross, J., Dingwall, J., and Dinh, H. (2014). *An Overview of Collaborative Contracting: Making Collaboration Effective and Choosing the Right Framework*. Melbourne: PCI Group.

Ruijter, H., van Marrewijk, A., Veenswijk, M. et al. (2021). 'Filling the mattress': trust development in the governance of infrastructure megaprojects. *International Journal of Project Management* 39 (4): 351–364.

Schein, E.H. (1985). *Organizational Culture and Leadership*. San Francisco, CA: Jossey-Bass.

Schein, E.H. (2004). *Organizational Culture and Leadership*. 3e. San Francisco, CA: Jossey-Bass.

Schön, D. (1975). Deutero-learning in organizations: learning for increased effectiveness. *Organizational Dynamics* 4 (1): 2–16.

Senge, P.M. (1992). *The Fifth Discipline: The Art and Practice of the Learning Organization*. New York: Doubleday/Currency.

Smircich, L. (1983). Concepts of culture and organizational analysis. *Administrative Science Quarterly* 28 (3): 339–358.

Transport (2011). *National Alliance Contracting Policy Principles*. Canberra: Department of Infrastructure and Transport.

Walker, D.H.T. and Lloyd-Walker, B.M. (2020a). Knowledge, skills, attributes and experience (KSAE) for IPD-alliancing task motivation. In: *Routledge Handbook of Integrated Project Delivery* (ed. D.H.T. Walker and S. Rowlinson), 219–244. Abingdon, Oxon: Routledge.

Walker, D.H.T. and Lloyd-Walker, B.M. (2020b). Behavioural elements of the IPD collaboration framework. In: *Routledge Handbook of Integrated Project Delivery* (ed. D.H.T. Walker and S. Rowlinson), 315–344. Abingdon, Oxon: Routledge.

Walker, D.H.T. and McCann, A. (2020). IPD and TOC development. In: *Routledge Handbook of Integrated Project Delivery* (ed. D.H.T. Walker and S. Rowlinson), 581–604. Abingdon, Oxon: Routledge.

Walker, D.H.T. and Rowlinson, S. (2020). IPD from a culture perspective. In: *Routledge Handbook of Integrated Project Delivery* (ed. D.H.T. Walker and S. Rowlinson), 197–218. Abingdon, Oxon: Routledge.

Widler, A. (2001). What anchors cultural practices? In: *The Practice Turn in Contemporary Theory* (ed. K. Knorr Cetina and E. von Savigny), 74–92. New York: Routledge.

Ybema, S., Yanow, D., Wels, H. et al. (2009). *Organizational Ethnography: Studying the Complexities of Everyday Life*. London: Sage Publications.

Zhang, L., Yao, Y., and Yiu, T.W. (2020). Job burnout of construction project managers: exploring the consequences of regulating emotions in workplace. *Journal of Construction Engineering and Management* 146 (10): 4020117.

2

Organising Occupational Health, Safety, and Well-Being in Construction: Working to Rule or Working Towards Well-Being?

Jing Xu and Yanga Wu

The Bartlett School of Sustainable Construction, University College London, UK

2.1 Introduction

The construction industry is a major employer. For example, in the UK, it employs 7% of the workforce. The industry is historically known for being unsafe, unhealthy, and bad for worker well-being. The fatality rate remains three times the industry rate for the last 40 years (HSE 2020). By 2021, there were 74 000 construction workers suffering from work-related ill health, mostly due to musculoskeletal disorders, stress, depression, and anxiety (HSE 2020). During 2011 and 2015, the UK construction industry represented 13.2% of the total number of in-work suicides, and the risk of suicide among construction male workers was three times higher than the male national average (ONS 2017). These statistics cover both office and site workers. A similar pattern is found in Australia, the US, and New Zealand (Milner 2016; Bryson and Duncan 2018; NIOSH 2021).

Ill health and injuries make working into later life difficult and foreshorten the working lives of construction workers (Ajslev et al. 2013). They also bring about a substantial cost to society and organisations. It is estimated that construction-related cancers cost UK society £189 million per annum (HSE 2016). Other occupational ill health costs construction employers £848 million each year (Gibb et al. 2018). It is notable that this figure is likely to be a substantial underestimate due to the long latency of some health conditions and the impact on the life experience of sufferers and their families.

Governments, client organisations, and large construction contractors have sought to address health and safety (H&S) issues at institutional, industrial, and organisational levels. The industry has experienced significant safety improvement after the introduction of H&S legislation, regulations, and standards. For example, in the UK, the Construction Design and Management Regulations clarify the H&S roles and duties of clients, designers, and contractors in the whole construction development process. Various H&S standards, such as ISO 45001, were developed to guide the safety systems design. In response, construction firms and major projects have devised and implemented H&S management systems, procedures, and behavioural programmes (Roberts et al. 2012; Lingard and

Construction Project Organising, First Edition. Edited by Simon Addyman and Hedley Smyth.

Turner 2017; Jones et al. 2019). Yet, the shortcomings of this legislation-driven approach became apparent in a fast-changing era as technological, economic, and other changes often outpace legislative changes (Ju and Rowlinson 2020). It helps grow a compliance culture that overemphasises formal routines and practices and encourages blame for human errors (Roberts et al. 2012). The overall consequence is the plateaued H&S performance and the reluctance to voluntarily invest for continuous improvement among construction organisations in the UK and many developed countries (Walters and James 2011; Chen et al. 2018; Smyth et al. 2019).

Meanwhile, well-being is rising up the construction agenda. It is suggested that construction workers have poor work–life balance, low sense of professional worth, and lack of job security (Sang et al. 2009), which plagues their well-being. Improving well-being will have a positive effect on performance but also help alleviate problems associated with the ageing workforce and skill shortages in a post-pandemic era and in a turbulent international labour market (Smyth et al. 2019; Jones et al. 2020).

This chapter aims to discuss how construction organisations can nurture the occupational health and well-being (OHW) of their people and break through the safety performance plateau. It draws on the findings of two research projects investigating the culture and digitalisation of OHW and safety in the UK construction industry (cf. Smyth et al. 2019; Duryan et al. 2021; Xu et al. 2021). Specifically, this chapter explores how OHW can be embedded in construction practices from the perspective of structuration theory (Giddens 1984). It contributes to project organising studies by explicating how and why the current OHW practices fail and by providing a strategic approach that can help embed OHW in practices and, hence, improve organisational performance.

2.2 Safety Management: A Tale of Two Paradigms

There are two distinct paradigms that influence safety management approaches. The old paradigm regards humans as the potential safety problem and tends to focus on human behaviour and the attributing factors (Dekker et al. 2007; Hollnagel 2014). Principally, safety rules, including safety policies and processes, are established by 'experts' (regulators, top management, or safety professionals). They are imposed on operatives as requirements that must be followed to avoid safety incidents without explaining and questioning the suitability of the rules to the unfolding situation. Non-compliances, regardless of the context, are seen as negative actions to be suppressed and corrected. Violations that contribute to accidents result in more extensive rules to prevent recurrence. The old view does not systematically and explicitly consider the evolving of safety rules in varying conditions. It sees rules as essentially top-down, static and in a sense linear (Hale and Borys 2013; Smyth et al. 2019). Site operatives are treated as the end users of safety rules with less competence and experiences than experts. Yet, they often assume the major responsibility of safety events as their violations and errors are perceived as the primary cause of safety incidents. Individuals may develop a defensive behaviour towards incident reporting and accident investigation due to the fear of being blamed, which reduces the reliability of reports and, hence, the rules derived from the evaluation process.

The new paradigm considers human error as a result of organisation-wide behaviour such as conflicting goals and time pressure, instead of the primary cause of safety incidents (Dekker et al. 2007). It treats humans as assets and puts human capital at the centre of creating and recreating rules (Hollnagel 2014). People enact rules, procedures, and checklists in their daily work; these acts generate patterns of action that accomplish organisational tasks (Feldman 2004). Therefore, it is important to incorporate operatives' experience into rule design so that processes can become routinised within the unit or division. This view is essentially bottom-up and dynamic. It regards 'violation', driven by human adaption and improvision, as inevitable and valuable to achieving safety outcomes (Hollnagel 2014), which should be managed rather than suppressed.

Construction projects are dynamic, and their successful management depends on the adaptability and improvisation of individuals (Olde Scholtenhuis and Dorée 2014). Skilled adaptions shape informal norms and practices that 'deviate' from the written rules, but they are essential to achieving performance outcomes. Yet, the continued use of informal norms and practices can result in inconsistent behaviour and outcomes across different groups within construction organisations. This points to the importance of identifying the gap between formal rules and those practices that lead to deviations at the firm level. According to Duryan et al. (2020), social learning and communities of practice are means for establishing internal and external synergies of rules. It also requires knowledge management (KM) systems to transfer knowledge across projects and organisational boundaries (Duryan et al. 2020). Consistent service and performance require a two-way process that establishes regular dialogue with operatives so that rules are continuously monitored and recreated (Xu et al. 2021).

2.3 Working Towards Occupational Health and Well-Being in Construction

The previous sections show that safety management practices in construction have been largely shaped by rules imposed by legislation, standards, and management systems. In contrast, the need for managing OHW emerges mostly bottom-up, with the growing awareness of health issues. The industry's efforts to balance H&S have been demonstrated by some major projects in the public domain (Jones et al. 2019), for instance, the London 2012 Olympic Park construction projects promoting 'health like safety' (Waite 2012). Healthy diet, exercise, meditation, and lifestyles have become a new focus of large construction organisations (Sherratt 2018).

Despite these good intentions, the effectiveness of such interventions has been limited (Lingard and Turner 2015). There are several reasons for this. First, these interventions tend to focus only on the immediate health risks as necessary for safety (Jones et al. 2019). The long-term effect of ill health is largely neglected. Second, OHW is managed too far down the hierarchy of control, mostly at site level through facilities and welfare. OHW issues are not treated in the same manner as safety to influence management decisions (Jones et al. 2019). There is an emphasis upon information sharing through safety management systems rather than knowledge sharing and transfer pan-project (Smyth et al. 2019). Some organisations used well-being initiatives as a corporate propaganda to support the brand and increase organisational competitiveness during the bid stage (Sherratt 2016).

Construction workers may perceive that OHW is a low priority compared to safety and other performance goals. They also lack OHW knowledge or good practice to positively influence their behaviour (Jones et al. 2019).

More importantly, OHW promotion programmes and events only address worker's unhealthy behaviour, which tends to be the symptoms of more fundamental causes in wider systems (Sherratt 2018; Hanna and Markham 2019). For example, the competitive nature of the industry sparked construction organisations to lower the labour cost and to operate with obstructive time constraints (Zou and Sunindijo 2015), which leads to a transient workforce, working long hours, unsocial work patterns, and abnormal shifts (Love et al. 2010; Smyth et al. 2019). The prevalence of subcontracting and self-employment makes it challenging to implement OHW policies and initiatives across projects and along the supply chains (Jones et al. 2019). The overall consequence of this context is a pressured work environment that can negatively impact OHW. Only focusing on superficial symptoms tends to draw attention away from more systemic and rooted problems and leaves the fundamental conflicts between construction practices and OHW intact.

2.3.1 The Structuration of Occupational Health and Well-Being

Embedding OHW means that OHW is structured in rules that enable and constrain actions in construction organisations and projects. The structuration process has two dimensions: the dimension of structure and the dimension of action (Giddens 1984). Specifically, *structures of signification* are rules that lend shared meanings of actions in particular contexts and guide the process of sense making and communication (Staber and Sydow 2002). For example, OHW is a part of organisational values, and workers have a shared understanding of what it is in day-to-day work. *Structures of legitimation* are rules that legitimise what and how actors should do in a particular context (Staber and Sydow 2002), such as regarding working long hours as 'the way we do things around here'. *Structures of domination* are means of production that actors use to actualise their purposes, which can be the controlling of materials and humans.

On the other hand, the dimension of action, which includes *communication*, *norms*, and *power*, shapes practices as observable expressions of structures (Giddens 1984). Giddens (1984) regarded agency not as intentions or motivations within individuals, but as the capability of acting otherwise. The implication for organising OHW is that knowledgeable workers are capable of choosing among rules and resources and integrating the contextual experiences to communicate meanings, form norms, and deploy power, in order to complete tasks and ensure safety and OHW. OHW structures provide the good conditions for the OHW of construction workers. Workers can reproduce structures by drawing upon pre-existing rules and resources to act, but they can transform the setting through flows of actions (Manning 2008). This calls for the management of interactions to sustain and develop OHW practices.

2.4 Methods

This chapter is based on data collected in two research projects on occupational health, safety and well-being (OHSW) in construction. Both the projects focused on construction organisations as the permanent firms in the project organising. The first project investigated the OHSW systems and cultures, and the second examined the digitisation for OHSW management. How to improve OHW is a topic shared by both the projects, nevertheless investigated from different perspectives. Both the projects used semi-structured interviews as the primary method to collect data and involved six types of organisations, including institutions, professional bodies, clients, designers, main contractors, and subcontractors. The interviewees included roles such as heads of H&S, site supervisors, human resource managers, and chief executive officers (CEOs). The overall sample represented a holistic and diverse view on OHW management in construction organisations.

An interpretative approach was taken in the data analysis to explore how OHW can be embedded in the construction practices. Data analysis was based on the constant comparing between theories and empirical findings and systematically combining themes (Dubois and Gadde 2002). The study draws upon structuration theory (Giddens 1984) as the framework of reference to look into whether and how OHW is structured as rules of signification and legitimation in daily work and how resources are allocated for OHW purposes in projects and organisations. Data extracted from the digitalisation project were analysed first, and themes emerged inductively to reflect the top-down and bottom-up processes of structuring OHW in construction practices, such as 'The meaning of OHW in management' and 'Interpretations of OHW'. The themes developed were then used to code data from the other project, adding extra themes where appropriate. Finally, the themes were reviewed, reorganised, and combined to finalise the main themes across the two projects.

2.5 Findings

This section presents the findings and shows the difficulties to embed OHW in construction practices as the top-down rules and the bottom-up actions were not aligned for improving OHW in construction.

2.5.1 Signification and Communication of OHW

2.5.1.1 Interpretations of OHW

Occupational health is commonly understood differently from safety as being the harm and diseases that workers get from the working environment.

> *Safety is the patent defects; it is an immediate thing. And health is the latent defect; it is the long term.* (CEO, Main Contractor 1)

Despite the difference, the industry has been addressing occupational health along with safety, which is the immediate effect. Construction employers generally lack the responsibility and accountability to deal with the long-term health risks both on site and in the office.

While the meaning of occupational H&S is well shared by workers, well-being is not. Most interviewees recognised well-being as a feeling, such as '*happiness*', '*feel loved*', and '*feel being looked after*', that is significantly influenced by the work environment. In this vein, improving well-being is to improve how workers are treated in the workplace. Respect and care were regarded as essential to well-being.

> *I think the whole is that they have to feel valued. . .I think wellbeing is about getting these basics right. . .I think it's that caring and listening.*
>
> *(Head of Commercial, Infrastructure Client 1)*

For site workers, the 'basics' also include welfare facilities such as warm site cabins, canteen, toilets, and shower. A site manager suggests that construction needs its own specific definition of well-being, given the work environment and tasks undertaken.

It was found that learning and personal development contribute to positive feelings in the workplace. Some large construction firms offer workers on-the-job training, apprentice programmes, and further education.

A rather different view is that well-being is the wholeness of life, for example,

> *People usually talk about work–life balance, to me it is a life balance. . .Well-being is having a peace of mind that your life balance around both physical and mental activities is aligned.*
>
> *(Group Head of Health and Safety, Real Estate Developer)*

Therefore, to improve workers well-being requires a good understanding of their demographics to ensure an appropriate fit between OHW initiatives and workers' needs. The interviews showed that, despite the diverse interpretations of what well-being means, the management process of caring is critical to the OHW of construction workers on site and in office.

2.5.1.2 The Meaning of OHW in Management

There is a lack of clarity in management initiatives, campaigns, and contracts in terms of what well-being means in construction organisations and on construction sites. Office workers were more comfortable talking about 'mindfulness', whereas site workers mainly discussed H&S operations and welfare when they were asked about well-being. The intersectional sphere between the workplace and the 'private', notably gender, family structures, and care responsibilities, receives little consideration in the current OHW initiatives and interventions. There is also uncertainty among workers as to how far employers are legitimised to invade personal life. 'Well-being' is most often used as an alternative to 'mental health'; the former is perceived as more acceptable and positive in communication than mental health. The lack of shared understanding discourages the open communication about OHW concerns and the adoption of OHW services.

2.5.2 Domination and Responsibility of OHW

2.5.2.1 The Tensions Between Business Priorities and OHW

In terms of resource allocation for safety and OHW, priority is still given to safety. A site manager commented that there is a limit on investment in OHW facilities and events on site, which needs justification with regard to its value for money. Money for safety is spent with less questioning and justification required. Yet, it was also mentioned, especially by site operatives, that financial criteria, hence costs and time, remain the top priority despite the espoused values and rhetoric about safety being a top priority. It was reported by both the main contractors and subcontractors that OHW initiatives and interventions can lead to increased costs. Justifying the value for money needs well-designed OHW systems, processes, and measures for outcomes, which requires more investment. However, the competitive bidding process drives construction firms to keep investment and expenditure low in order to secure contracts. Hence, contractors had no internal incentive to make fundamental changes, unless it was required by client organisations, and OHW costs were taken into account in the tendering process.

The investment in OHW tends to be tactical, providing fruits, wristbands, and gym membership for example, rather than strategically as part of the business model and capability development. OHW issues are managed at the lower level of hierarchies and are not integrated with safety and other parts of the project business. The working-to-rule approach that helped safety is unlikely to work for OHW as OHW processes are not regulated, nor is it supported by organisational systems to coordinate organisational capabilities for OHW at the project level. Indeed, the successful history of safety management was found to hinder the further improvement for safety and OHW.

> *At this stage I think the majority of the construction businesses do have comfortable systems from the past, which they enjoy working with, and also they enjoy blaming for, but no one is willing to make the change. It's just because they feel complacent, they feel comfortable...The mindset unfortunately is not yet switched on, because what happens is sometimes there is a job protection, just job justification.*
>
> *(H&S Manager, Main Contractor 2)*

Changing managements' mindset and systems to go beyond compliance is very much needed for OHW given that the majority of senior and middle managers did not consider well-being as a major responsibility of employers. There was only one senior manager from an international construction firm who believed that '*Responsibility of wellbeing is no difference of managing any risks*' (H&S Director, Main Contractor 3).

2.5.2.2 Imbalanced Power and Responsibility Between Management and Workers

The responsibility is then shifted to individuals, with organisations only taking a supportive role. For example, most organisations have mental health first-aider training programmes for line managers. Although organisations send regular messages about cancer, stress, smoking, and other ill health, it is the individuals that are responsible for transferring OHW knowledge and applying it in practice. Despite the overreliance on individuals, frontline managers and operatives are rarely involved in the decision-making on OHW issues. Many prescriptions, initiatives, and procedures are developed by 'university' educated office staff, who have little or no understanding of how operations are or can be conducted safely and healthily on site. Some equipment and procedures can actually make certain activities riskier. Goggles were given as an example, restricting vision for important cutting tasks and inducing headaches. Furthermore, some subcontractors pointed out that large clients and main contractors tend to impose H&S policies down the supply chain, which can disrupt their own policies and efforts. Well-being policies could prove more disruptive due to the divergent interpretations of well-being in and across organisations in the supply chain.

The lack of trust in the workforce competence was reported as one of the main reasons for the low level of workforce involvement in the decision-making. Another reason is the prevalence of subcontracting and self-employment in the construction industry.

> *I think the culture of zero hours contracts and short-term contracts coming into a job mean there's no real loyalty from staff or to staff. . .There's not that level of empowerment to allow people to find their way and understand the task better. . .I don't think there's the mindset to change the culture in construction to have more understanding of the wants and needs of individuals. I think it's just a resource heavy: we need that to be done, it's got to be done for then, so get a team of 10 guys to do it. If two of them don't work, we'll get another two guys in.*
>
> *(Deputy Chief Executive, H&S Professional Body)*

Overall, the power between organisations and the workforce is imbalanced. Workers generally lack the sense of ownership of OHW initiatives, which are not consistently supported by organisational systems and resources.

2.5.3 Legitimation, Norms, and OHW

2.5.3.1 'Norms' of Construction Project Business and Works

Competition, efficiency, flexibility, and personal accountability were recognised as the legitimised rules of working in the construction industry. The legitimation is reinforced by practices such as competitive tendering, payment by results, subcontracting, temporary contracts, and self-employment. As a consequence, long working hours, irregular shifts, and working at weekends for meeting deadlines and long commutes are regarded as normal conditions. Organisations encourage this type of behaviour through bonuses and other rewards, reinforcing the 'working equals earning' mentality of the industry.

Some contract operatives interviewed were less concerned about OHW; they are motivated by working intensively and extensively so they could move to the next job and maximise their income. Other workers see the harm to mental health, such as anxiety due to the tight deadline and stress due to the difficulties in sustaining interpersonal relationships. It was raised that mental health problems can further lead to H&S risks. For example, many office and site workers reported that fatigue is a major risk on site, when travelling to and from work and can lead to poor decision-making. Furthermore, working long hours under fatigue incentivises drug use to stay awake. Despite this, most workers accept the conditions as it is how the construction project business works.

2.5.3.2 The Influence of 'Norms' on Trust and OHW

The existing OHW initiatives and interventions were found to be not well adopted and embedded in routines, especially by site operatives. One reason reported is that OHW initiatives can conflict with commercial interests that are perceived as the top priority by some site operatives, particularly contract workers. There is insufficient cost being built into bid prices to afford the time for OHW-related activities. Also, terms of pay can incentivise operatives to work during weekends when the opportunity arises, which brings in needed income but considerably disrupts home life. Another reason is the low level of trust between operatives and management. For example, some digital technologies for OHW pose problems as intrusive interventions. They can be perceived, indeed used, as monitoring devices by management that are unwelcome by operatives and lead to pressures to intensify work rates that induce stress and fatigue.

This type of action erodes trust and feeds perceptions of suspicion as to the motivations of management. The situation is further worsened by the blame culture and the prevalence of temporary contracts that make workers more vulnerable. There is resistance to reporting OHW issues and using mental health support services. Apart from trust issues, operatives were aware that the information will not be analysed and acted on in some cases. Compliance was perceived as the driver to practical support, rather than care for people, as commented by an H&S manager:

> *The modern way of thinking is actually towards the human rather than towards the plant and equipment, because with safety now we can control it. How do we control health? Very difficult. How do we control wellness, or not only control, make it better? Again, difficult.*
>
> *(H&S Manager, Main Contractor 2)*

Last but not least, the masculine norms such as competitive ethos and stoicism pertaining to illness and injury were found to negatively influence the OHW of construction workers. It is clear that masculinity at work potentially masks emotions, health issues, and expressions of vulnerability through pressure and coercion:

> *It's very gung-ho get the job done. It's very macho. We've got guys out there who will just [be] solider on to get the job done and if somebody is not feeling well one day, they'll get ridiculed.*
>
> *(H&S Professional Body, Deputy Chief Executive)*

The macho culture and male-dominated workforce profile also influences the OHW of women workers. It was mentioned that few OHW initiatives consider the influences of gender-specific issues such as caring responsibilities on OHW, safety, and job performance, resulting in inequality to OHW services and promotion. The latter can further damage the mental health of women workers.

2.6 Discussion

The empirical findings highlight the failure to embed OHW practices into the organisation and project routines suitable for the operational context of each provider, especially at subcontractor level. The failure is mainly due to (i) the lack of management process of caring and shared understanding of what well-being means in construction practices, (ii) the hierarchical and silo way of managing OHW and thus the lack of ownership of OHW by operatives, and (iii) the transactional business model and masculine norms among construction organisations, compounded by the low level of trust between the management and workforce. This section further discusses the findings and suggests mechanisms and conditions that can help the structuration of OHW in construction practices.

The interview data showed that a sense of caring is central to the well-being of office and site workers in construction. The management processes that generate this sense go beyond compliance to rules that focus on organisation. Rather, demonstrating the sense of caring requires practices that depend on relationships, or the 'basics' as trust and respect, and focus on organising and continual enactment for effective implementation. It is through the ongoing process through routines that construction workers comprehend, reproduce, or refine the meaning of well-being in daily operations. Extant studies have pointed out that current practices that address OHW along with safety or through public health measures are insufficient (Sherratt 2016), for it does not deal with the long-term effects of ill health (Jones et al. 2019). This study further argues that the hierarchical and prescriptive approaches are ineffective as they fail to generate the sense of caring through implementing initiatives. Such approaches assume that the one-size-fits-all measures, such as diet and lifestyle, are needed by all construction workers in the same manner, which is clearly not the case. The findings pointed out that there are more dimensions that deserve consideration in OHW initiatives in construction, notably gender, family structures, and education. Management needs a better understanding of the workforce in order to specify the meaning of OHW and design organisational initiatives.

Embedding the ethics of care in the management process is more than a disposition. It is a social practice constructed in relationships and interactions. Bottom-up activities are imperative to embedding OHW in construction practices. They induce bridges between individuals and the organisation, helping build a better understanding of the workforce. They are also mechanisms that empower the workforce and supply chains to share their knowledge and experiences, become involved in the decision-making, and recreate practices that are suitable to the operational context and enhance skilful performance. Yet, the findings demonstrated that bottom-up processes are lacking in most construction organisations. This is less about communication procedures and technologies than the practices of sharing and transferring knowledge.

Strategic investment in management capability is a key driver to incrementally transforming practices (Smyth 2022). Previous studies have stressed the importance of human resource management (HRM) and KM to OHW in project contexts (Turner et al. 2008; Asquin et al. 2010; Duryan et al. 2020). HRM and KM stress the importance of developing human capital to well-being (Turner et al. 2008; Duryan et al. 2020). This study revealed that effective organising of human capital depends on the social capital of construction organisations. Social capital entails networks of relationships that include norms, values, and obligations, which needs investment in relationships (Coleman 1988). This points to the significance of relationship management to help systems integration through nurturing relational norms such as trust and respect within and between organisations. It is the integrated systems at meso- and macro-level and enhanced human and social capital that ensures behavioural consistency beyond compliance.

More fundamentally, a transformational business model is needed to break down the silos and increase integration. The current transactional business model of construction organisations emphasises cash flow management and the return on capital employed, instead of organisational capability development and long-term growth (Smyth 2022). The resulting practices of prioritising commercial interests in operations are conflicted with OHW initiatives, rendering difficulties to embed OHW into daily practices. To a broader view, the dominant business model needs change as the increasing complexity of construction projects and service demands have outstripped the capabilities of construction organisations to meet the demands (Smyth 2022). People are the most important ingredient in the changing landscape of competition based on service differentiation and value delivery, and, arguably, their well-being is fundamental. A transformational business model prioritises performance improvement across a broader range of criteria. It shifts the focus towards an investment-led and people-centred approach to enhance management and technical capabilities to improve productivity, service provision, and value outcomes. It also engenders more bottom-up activities that balance the responsibility for OHW between the organisation and individuals. The sense of caring can be created and flowed from this process, and hence, OHW improved.

Current industry norms place considerable constraints in improving OHW. Shifting from a blame and macho culture to a just and caring culture is found to be necessary for improving OHW in construction. A just culture prioritises learning and human adaptability over blame, and it does not tolerate gross negligence, wilful violations, and destructive acts (Dekker 2007). In construction organisations, a just culture can foster psychological safety, trust, and norms of care and respect, an atmosphere that reduces the perception of invasiveness and intrusiveness and thus encourages workers to adopt OHW initiatives. The macho culture of the construction industry is not only due to the historically few, albeit changing profile of women workers but also due to the neoliberal policies and practices that legitimise a set ideology, such as efficiency, competition, flexibility, and personal accountability (Hanna et al. 2020). In other words, promoting OHW in the construction industry is largely constrained by the industry structure and action (Hanna and Markham 2019). This points to the need to critically review the viability of the neoliberal working landscape, to create awareness of need for change across industry ecosystem, and to reshape working practices for workplace equality to OHW.

2.7 Conclusion

OHW is a work in progress. The construction industry has greatly reduced its fatalities and injuries through legislation and compliance, a reactive and top-down approach aligned with the old paradigm of safety management. Recently, more proactive management approaches, such as high reliability organising, have been promoted (Xu et al. 2021), reflecting the evolving to the new paradigm that values human knowledge, interactions, and systems thinking.

OHW has historically not been given the same precedence that safety has within the construction industry, and the recent initiative of managing health like safety has received limited impacts. A major reason is that most interventions only treat the symptoms of poor

OHW, rather than the fundamental causes. Enhancing OHW requires a strategic approach that embeds OHW in construction practices. Based on the interview data from two research projects and taking structuration as a viewpoint, the findings highlight the failure to embed OHW in construction practices and identify mechanisms that help the embedding process. It is suggested that construction organisations need a better understanding of the workforce and more bottom-up activities in the creation and refinement of OHW initiatives, focusing on the organising processes of balanced power and responsibility, rather than the prescription and imposition of accountability.

More importantly, the transactional business model needs change to incrementally invest in organisational capabilities to develop social and human capital. Particularly, this chapter stresses that relationship management is a critical part of structuring for OHW as it helps break down the silos in construction organisations and improves system integration. Lastly, the transformational business model is accompanied by a just and caring culture to encourage open communication and knowledge sharing of OHW issues, nurture norms of trust and respect, and enhance equality to OHW in the construction industry.

To further enhance OHW in construction, future research should devote more attention to unearthing specific organisational practices, and specifically routines, that can enhance OHW, particularly how the ethics of care can be created and recreated in the management process. It points to the significant role of strategic leadership within construction organisations to transform the business model, invest in management capabilities, and hence deal with the root causes of poor OHW in construction.

References

Ajslev, J.Z.N., Lund, H.L., Møller, J.L. et al. (2013). Habituating pain: questioning pain and physical strain as inextricable conditions in the construction industry. *Nordic Journal of Working Life Studies* 3 (3): 195–128.

Asquin, A., Garel, G., and Picq, T. (2010). When project-based management causes distress at work. *International Journal of Project Management* 28 (2): 166–172.

Bryson, R. and Duncan, A. (2018). *Mental Health in the Construction Industry Scoping Study*. NZ: Branz www.branz.co.nz/pubs/research-reports/sr411.

Chen, Y., McCabe, B., and Hyatt, D. (2018). A resilience safety climate model predicting construction safety performance. *Safety Science* 109: 434–445.

Coleman, J.S. (1988). Social capital in the creation of human capital. *American Journal of Sociology* 94: 95–S120.

Dekker, S. (2007). *Just Culture: Balancing Safety and Accountability*. Ashgate.

Dekker, S., Siegenthaler, D., and Laursen, T. (2007). Six stages to the new view of human error. *Safety Science Monitor* 11 (1): 1–5.

Dubois, A. and Gadde, L.-E. (2002). Systematic combining: an abductive approach to case research. *Journal of Business Research* 55 (7): 553–560.

Duryan, M., Smyth, H., Roberts, A. et al. (2020). Knowledge transfer for occupational health and safety: cultivating health and safety learning culture in construction firms. *Accident Analysis and Prevention* 139: 105496.

Duryan, M., Xu, J. and Smyth, H., (2021). Cultivating a "just" culture in construction industry to improve Occupational Health and Safety management systems: lessons learnt from aviation. In: *CIB W099 & W123 Annual International Conference: Changes and innovations for improved wellbeing*. Glasgow.

Feldman, M.S. (2004). Resources in emerging structures and processes of change. *Organization Science* 15 (3): 295–309.

Gibb, A., Drake, C., and Jones, W. (2018). *Costs of Occupational Ill-Health in Construction*. UK: Institute of Civil Engineers.

Giddens, A. (1984). *The Constitution of Society: Outline of the Theory of Structuration*. Cambridge: Polity Press.

Hale, A. and Borys, D. (2013). Working to rule, or working safely? Part 1: A state of the art review. *Safety Science* 55: 207–221.

Hanna, E.S. and Markham, S. (2019). Constructing better health and wellbeing? Understanding structural constraints on promoting health and wellbeing in the UK construction industry. *International Journal of Workplace Health Management* 12 (3): 146–159.

Hanna, E., Gough, B., and Markham, S. (2020). Masculinities in the construction industry: A double-edged sword for health and wellbeing? *Gender, Work & Organization* 27 (4): 632–646.

Hollnagel, E. (2014). *Safety-I and Safety-II: The Past and Future of Safety Management*. New York: Taylor & Francis.

HSE (2016). *Costs to Britain of Work-Related Cancer*. UK: Health and Safety Executive.

HSE (2020). *Construction Statistics in Great Britain, 2020*. UK: Health and Safety Executive.

Jones, W., Gibb, A., Haslam, R., and Dainty, A. (2019). Work-related ill-health in construction: the importance of scope, ownership and understanding. *Safety Science* 120: 538–550.

Jones, W., Chow, V., and Gibb, A. (2020). *Covid-19 and Construction: Early Lessons for a New Normal?* Loughborough University.

Ju, C. and Rowlinson, S. (2020). The evolution of safety legislation in Hong Kong: actors, structures and institutions. *Safety Science* 124: 104606.

Lingard, H. and Turner, M. (2015). Improving the health of male, blue collar construction workers: a social ecological perspective. *Construction Management and Economics* 33 (1): 18–34.

Lingard, H. and Turner, M. (2017). Promoting construction workers' health: a multi-level system perspective. *Construction Management and Economics* 35 (5): 239–253.

Love, P.E.D., Edwards, D.J., and Irani, Z. (2010). Work stress, support, and mental health in construction. *Journal of Construction Engineering and Management* 136 (6): 650–658.

Manning, S. (2008). Embedding projects in multiple contexts – a structuration perspective. *International Journal of Project Management* 26 (1): 30–37.

Milner, A. (2016). Suicide in the Construction Industry: Report by Deakin University for MATES.

NIOSH (2021). *Construction Safety and Health*. UK: National Institute for Occupational Health and Safety.

Olde Scholtenhuis, L.L. and Dorée, A.G. (2014). High reliability organizing at the boundary of the CM domain. *Construction Management and Economics* 32 (7–8): 658–664.

ONS (2017). *Suicide by Occupation, England: 2011 to 2015*. UK: Office for National Statistics.

Roberts, A., Kelsey, J., Smyth, H. et al. (2012). Health and safety maturity in project business cultures. *International Journal of Managing Projects in Business* 5 (4): 776–803.

Sang, K.J.C., Ison, S.G., and Dainty, A.R.J. (2009). The job satisfaction of UK architects and relationships with work-life balance and turnover intentions. *Engineering, Construction and Architectural Management* 16 (3): 288–300.

Sherratt, F. (2016). Shiny happy people? UK construction industry health: priorities, practice and public relations. In: *32nd Annual ARCOM Conference* (ed. P.W. Chan and C.J. Neilson), 487–496. Association of Researchers in Construction Management: Manchester, UK.

Sherratt, F. (2018). Shaping the discourse of worker health in the UK construction industry. *Construction Management and Economics* 36 (3): 141–152.

Smyth, H. (2022). Transforming the construction firm? In: *Describing Construction* (ed. J. Meikle and R. Best). Abingdon: Routledge.

Smyth, H., Roberts, A., Duryan, M., et al., (2019). Occupational Health, Safety and Wellbeing in Construction: Culture, Systems and Procedures in a Changing Environment. The Bartlett School of Construction & Project Management.

Staber, U. and Sydow, J. (2002). Organizational adaptive capacity: a structuration perspective. *Journal of Management Inquiry* 11 (4): 408–424.

Turner, R., Huemann, M., and Keegan, A. (2008). Human resource management in the project-oriented organization: employee well-being and ethical treatment. *International Journal of Project Management* 26 (5): 577–585.

Waite, L. (2012). In practice London 2012. *Perspectives in Public Health* 132 (2): 56.

Walters, D. and James, P. (2011). What motivates employers to establish preventive management arrangements within supply chains? *Safety Science* 49 (7): 988–994.

Xu, J., Duryan, M., and Smyth, H., (2021). Digitalisation for occupational health and safety in construction: A path to high reliability organising? CIB W099 & W123 Annual International Conference: Changes and innovations for improved wellbeing. Glasgow.

Zou, P. and Sunindijo, R. (2015). *Strategic Safety Management in Construction and Engineering*. Chichester: Wiley Blackwell.

Part II

Wider Integration

Although no project is independent of the organisations around it, they are often treated as such across a number of different practices. In this part, we bring together work that highlights just how much projects are integrated with the contexts in which they sit and how these wider contextual features influence organising.

Whyte and Davies take a system-of-systems perspective, arguing for an outcome-focused approach based on the premise that construction projects intervene, evolve, and are heterogeneous to recognise the range of complexity and uncertainty in the complex systems and outcomes in messy and uncertain contexts.

Coffman and Kelsey explore project finance drawing on economic theories of the firm to explore what the implications of financing models are for how projects might need organising, the conceptual analysis being supplemented by a case study.

Morgan et al. look at information technology platforms and modularity for the adoption of building information modelling (BIM). Their aim is to develop a future state perspective for understanding the transformation of the digitisation of the industry. This type of organising is to be welcomed as it continues to widen the perspective as to the issues involved in organising the adoption of BIM and other digital technologies, and points towards research and practice as the extent to which firms graft the technology into existing systems or whether systems need reorganising to optimise the technology benefits.

Klakegg and Olsson look at resilience in project organising. This links together two areas for the first time, which they do by looking into governance in terms of the flexibility and survival of projects.

3

Systems Integration in Construction: An Open-Ended Challenge for Project Organising

Jennifer Whyte[1] *and Andrew Davies*[2]

[1]*John Grill institute for Project Leadership, University of Sydney, Australia*
[2] *Science Policy Research Unit, University of Sussex Business School, Brighton, UK*

3.1 Introduction

Systems integration is a key issue in the delivery of projects. It is the process of making constituent parts of delivered systems (both diverse knowledge and physical components) work together (Whyte and Davies 2021) to achieve project outcomes. As projects become defined by their end results, focus must be given to the systems they deliver and how these are integrated to achieve outcomes.

There is growing recognition of the importance of focusing on systems integration in construction, with recent work promoting a systems approach (ICE 2020). As part of an Initiation Routemap, new guidance for project sponsors articulates how to set up projects for systems integration (IPA 2021). Systems integration has two faces (Hobday et al. 2005), involving work *upstream* with the supply chain to integrate sub-systems and systems (Gholz et al. 2018) and *downstream* with operators and end users as systems are delivered. Where systems cannot be fully specified and understood at the start of a project, systems integration requires a flexible and adaptive process to deal with emerging complexity and uncertainty (Whyte and Davies 2021). The purpose of a project is to create the preconditions for other activities (Smyth 2018). Thus, construction projects deliver value to their clients, end users, and societal stakeholders, not from inputs during delivery but as a built system in use post-completion within a wider set of systems.

As the construction sector seeks to transform itself, there is a need for specific attention to an outcome-focused approach to systems integration to develop a delivery model that delivers safety-critical systems, to manage supply chains to achieve net zero pollution, and to modularise production to benefit from advanced manufacturing approaches. We contend that systems integration is a particular challenge in achieving these goals, especially as new digital systems become connected to legacy analogue systems in infrastructure projects.

Systems integration methods aim to keep focus on the project outcomes or ends. This involves partitioning systems with a view to how they are going to be integrated, and to

Construction Project Organising, First Edition. Edited by Simon Addyman and Hedley Smyth.

collaborating – as a means, rather than an end in itself – with a view to how whole systems will be delivered to achieve project outcomes. We, therefore, focus on whole system outcomes that projects deliver. Those working on engineering systems increasingly recognise they are open and socially as well as technologically complex (De Weck et al. 2011; INCOSE 2014). Yet, in early organisational research, the sociologist Selznick (1953) studied the building work by the Tennessee Valley Authority, observing how means can easily become ends, with organisations set up for one purpose finding ways to perpetuate their existence beyond their original objectives. We, thus, argue for project managers, involved in construction project organising, to focus strongly on the project outcomes and the systems through which these are delivered (rather than means). Dominant interests in means hamper integration. This can take the form of a focus on time, cost, and quality within scope at project level, or on competitive tendering dominated by price at firm level, which leads to low levels of investment and capabilities.

We argue that an outcome-focused approach to systems integration has distinctive features in construction project organising for the three reasons that construction projects are: interventions, evolving, and heterogeneous.

1) First, since projects intervene in wider systems of systems, they have to be integrated across their boundaries into complex social, material, and environmental systems (Hughes 2004). Systems of systems are collections of systems: though each (designed) system is owned and managed separately, there are significant interdependencies with shared resources and emergent properties across these systems, which include naturally occurring systems as well as human-designed systems (Whyte et al. 2020).
2) Second, projects are a temporally changing, or evolving, phenomenon. Disciplines join and leave the project, and while work can (and should) be done to understand complexity and uncertainty at the outset, many projects are of significant scale, with emergent complexities and uncertainties that have to be addressed during the process of systems integration. This means that systems integration in construction cannot be addressed solely in terms of arriving at a fixed and stable system that may go through phases of planned change through the life of the project and into operations, but also requires attention to how the system is itself in the process of integrating. Identifying and managing the emerging complexities and uncertainties becomes key in managing systems integration (Whyte and Davies 2021).
3) Third, the sector is heterogeneous. Construction projects are inter-organisational in nature and their supply is characterised by the absence of large-scale corporations that claim systems integration as a core capability.

These three reasons highlight the inter-organisational nature of projects and the characteristics of supply in construction project organising, with the absence of large-scale corporations that claim systems integration as a core capability. In the next section, we outline these challenges of systems integration in construction. In the following section, we then describe the origins and history of systems integration, both in terms of practical developments and the associated theoretical work on technical coordination and cooperation. The subsequent section discusses systems integration in construction, elaborating on the contexts of use and their importance. In the final section, we draw conclusions and set out further areas for research.

3.2 Challenges of Systems Integration in Construction Project Organising

Systems integration has grown from its origins as a topic of concern in aerospace and defence into many sectors, such as telecoms and rail transportation. But these industries have large, integrated engineering companies performing systems integration. Construction is different because of the absence of these firms and the prevalence of highly specialised firms and disciplines. There have been some high-profile failures of large, complex construction projects to deliver integrated systems on time and on budget. Examples include Berlin Brandenburg airport, which was nearly a decade late, and also Crossrail, which spent many years more than planned in integrating, testing, and commissioning the rail assets it delivered. These high-profile failures have also led to wider interest in systems in the built environment, with recent work arguing for civil and infrastructure engineers to shift mindset to a systems approach (ICE 2020), setting out a research agenda (Whyte et al. 2020), and questioning the criteria for the successful delivery of infrastructure projects (Winch 1998; Gil and Fu 2022).

For more than two decades, The Netherlands has pioneered the implementation of systems engineering approaches to infrastructure, to manage critical water and transportation infrastructure (De Graaf et al. 2017). There is also work taking a systems approach to infrastructure at a national and regional level in the UK (Hall et al. 2013). The International Council on Systems Engineering has developed guidance on systems engineering in infrastructure (Kouassi 2015) and has worked with a European initiative on large infrastructure projects in Europe to apply this (Staal-Ong et al. 2016). Drawing on experience at the UK's Heathrow Terminal 5, a systems approach to construction projects has been advocated (Blockley and Godfrey 2017). Many complex projects in the sector now use the classic systems engineering 'V' diagram as a representation of the project delivery process, to emphasise the connection at different levels of decomposition and systems design of deliverables at programme, project and asset level, and the integration and testing of the built deliverables.

Yet, as outlined in the Introduction, three things make systems integration a particular challenge in construction.

3.2.1 Construction Projects as Interventions

First, construction projects are interventions into existing environments, both natural and built, and thus, the systems that they deliver are open ones, connecting into these environments (Hughes 2004). To revisit projects as interventions, Whyte et al. (2020) invert the familiar V cycle for project delivery to show the project as one of a number of interventions in natural and built systems (see Figure 3.1).

This complexity is physical (involving both technical and natural systems) and also organisational. Delivered through inter-organisational projects (Sydow and Braun 2018),

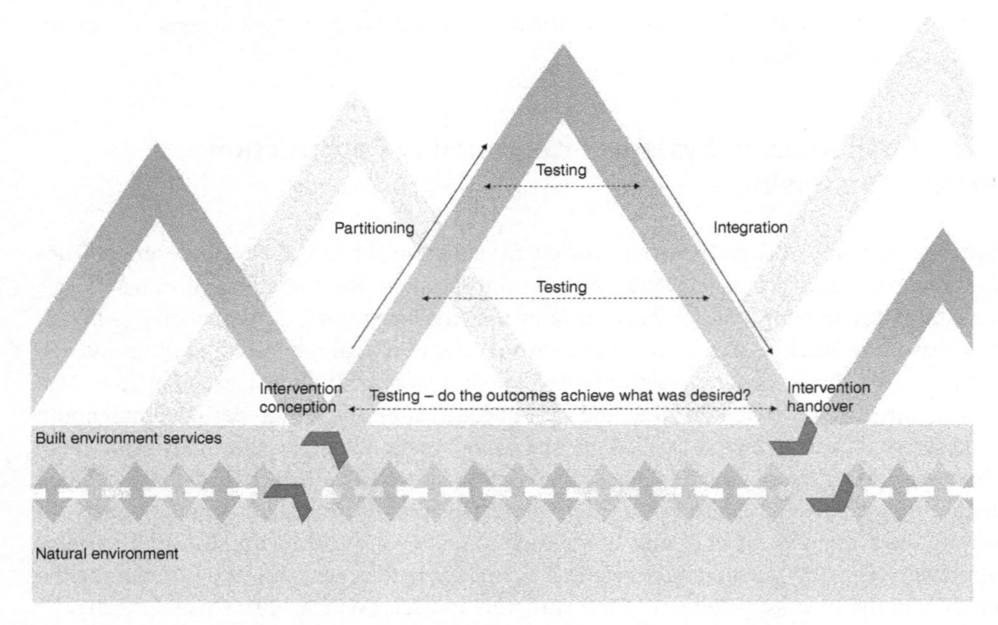

Figure 3.1 Inverting the classic 'V' diagram, projects are shown as interventions into the natural environment and built environment services provided by operating infrastructure *Source:* Whyte et al. (2020) / Taylor & Francis.

construction projects, thus, operate within a wider project ecology and have a range of technical and organisational interfaces.

3.2.2 Construction Projects as Evolving Phenomena

Second, projects grow and change shape, as individuals and organisations join and leave the effortful work of delivering a project as an emergent accomplishment. Professionals from different disciplines not only join and leave the project at different stages of delivery but also may have different lengths of commitment to the work of delivering the project. Over time, this has led suppliers to focus on tasking and finishing, and project managers to focus on managing through contracts, rather than integrated delivery. Understanding how to coordinate across disciplines is a growing challenge, particularly as project deliverables on such projects become increasingly cyber-physical, with sensors and control systems as well as physical components. While work can (and should) be done to understand complexity and uncertainty at the outset, many projects are of significant scale, with many emergent complexities and uncertainties.

An example of a source of complexity and uncertainty that affects construction is the relationship with the ground (Biersteker et al. 2021) – a relationship that can be analysed and understood, yet in many cities may also reveal surprises, such as a forgotten cable or water main, or an archaeological site. While firms play significant roles in systems integration in these project-based industries (Hobday et al. 2005), in parts of this landscape of construction project organising, there are no dominating firms (Rutten et al. 2009), and

Table 3.1 General systems integration concerns at different stages of project delivery in construction.

Project stage	Systems integration concerns
Set-up	Outcomes focus, projects as interventions, functional analysis, systems architecture, distribution of roles and responsibilities, governance of configuration management and interfaces, planning forward
Evolution	Maintaining outcomes focus, design change, scope creep, interfaces, supply chain integration
Transition to operations	Outcomes focus, testing, commissioning, handover, integration into contexts of use

projects may involve large-scale and long-term organising. Questions arise, therefore, about the distribution of accountability and responsibility for systems integration across the different parties – the owner-operator, project sponsor, delivery client, design consultants, contractors, and subcontractors.

Table 3.1 outlines some general concerns through the set-up, evolution, and transition to operations of construction projects, and we further expand on this in relation to building and infrastructure projects across the spectrum of complexity from systems and systems of systems below.

3.2.3 Construction Projects as Heterogeneous

Third, as a project-based industry, construction involves projects that vary considerably in size and nature, with different levels of internal complexity and externally as interventions into infrastructure (Whyte et al. 2019). Construction is a more heterogeneous sector than the classic manufacturing sectors. The projects delivered can be usefully characterised as follows:

1) Extremely complex projects (£10bn+) – these are large-scale programmes of work with multiple sets of projects (including examples such as High Speed 2 in the UK, which was budgeted at £45bn but increases in cost and is reduced in scope as complexity emerges). Such projects have considerable complexity at the systems of systems level, with significant political and market visibility and complexity.
2) Large projects (£1bn–£10bn) have high internal complexity of projects and sub-projects and may have visibility and external complexity.
3) Building and infrastructure projects (usually ~£1m–£999m) – most projects are smaller although potentially with non-standard requirements and high internal complexity.
4) A portfolio of infrastructure interventions that include the infrastructure owners' network maintenance projects in sectors, such as roads, water, and power, and the housing retrofit programme.

The degree of project complexity (Shenhar and Dvir 2007; Brady and Davies 2014) and number of interfaces to be managed will shape, and be shaped by, the approach to systems integration. A process approach to systems integration (Whyte and Davies 2021) suggests that while it can be designed to address known organisational and technological complexity and uncertainty, in extremely complex projects, it is not accomplished once, but rather that there need to be flexible and adaptive organising processes to deal with emerging complexities and uncertainties (both organisational and technological). The spectrum of types of projects that we distinguish is summarised in Table 3.2, showing how they span across building and infrastructure projects.

Such projects include those organised as a programme, for example infrastructure and cities (collections of interconnected infrastructure and buildings), and those organised as a project that delivers a single system, for example individual buildings, bridges, roads, water, power, and maintenance of systems.

On a systems project, the internal complexity of the project can be planned for and managed, and a lot of the emergent uncertainty may emerge across the boundary of the project with the contexts of use. At the technical level, each system is a system of products (which then need to be interfaced). Here, a concern of systems integration is to ensure the project delivers intended outcomes as an intervention into wider urban systems.

We have argued elsewhere that the client plays an important role in organising integration at the start of extremely complex projects (Whyte et al. forthcoming). These projects are characterised by a high degree of internal complexity, with systems integration activities at system and programme levels. While the client retains accountability for systems integration on such projects – and this may involve legal obligations in relation to a safety case and the appointment of a legally accountable person – a key strategic question the client faces is how to assign roles and responsibilities for systems integration at programme and technical levels across the client, supply chain of designers and contractors, delivery partner, chief engineer, and/or contracted systems integration firm. As the IPA (2021) project routemap outlines, this distribution of responsibilities, while retaining overall accountability, raises questions of governance and organisational design with each alternative raising risks that need managing throughout the process of delivery.

Table 3.2 A spectrum of types of projects.

Types of projects	**Extremely complex infrastructure projects (typically £10bn+)**	**Large infrastructure projects (typically £1bn–£10bn)**	**Building and infrastructure projects (typically £1m–£999m)**	**A portfolio of infrastructure interventions**
Systems integration	At meta-systems and systems level internally and across project boundaries		At systems level internally and across project boundaries	
Internal complexity	High, though modular methods may be used to reduce complexity		May be relatively low, and modular methods can be used to reduce complexity	
External complexity	High, though there may also be high internal complexity		Can be high relative to internal complexity	
Challenges of systems integration	Distributing responsibility while retaining accountability, oversight and management of interfaces with focus on technological and organisational complexity and uncertainty at system and programme levels and across boundaries.		Maybe organisationally simpler, but require oversight and management of technical interfaces with focus on complexity and uncertainty at systems level and across boundaries	

3.3 From the Origins of Systems Integration to Its Application in Construction

Systems integration did not originate in construction, but was developed through the work of the Ramo-Wooldridge Corporation on the post–Second World War Atlas missile project (Hughes 1998; Davies 2017). The below sub-sections consider the questions raised by the origins and development of systems integration, where an early framing as technical coordination raises questions about the separation of organisation and technology, and early theorising raises questions about the temporal nature of systems integration processes and the role of cooperation in achieving collaboration. We then consider the implications for systems integration in construction project organising.

3.3.1 Origins of Systems Integration

Systems integration has its origins as part of the technical discipline of systems engineering, which developed the processes and structures for integrating technological systems delivered through large, novel, complex projects. Systems engineering emerged alongside the adjacent disciplines of project management and operational research during the development of weapons and defence systems projects in the 1950s. As Johnson (2003) explains:

> *'The government, industry, and academia were all central to the creation of project management, and its affiliated methods, systems engineering and operations research. Project management primarily addressed the organizational issues, systems engineering the technical coordination, and operations research—which became systems analysis, the technical feasibility of projects'*
>
> *(Johnson 2003: p. 679)*

Many techniques and processes associated with systems integration emerged to address the challenges involved in the development of the Atlas Project, which developed the world's first intercontinental ballistic missiles (Hughes 1998; Johnson 2013).

The Ramo-Wooldridge Corporation was employed as the world's first dedicated 'system integrator' firm on the Atlas Project in 1954. It was responsible for coordinating '*the work of hundreds of contractors and development of thousands of sub-systems*' (Mahnken 2008, p. 38) and developing a broad base of capabilities in:

> *'guided missile research and development, aerodynamics and propulsion systems, communication systems, automation and data processing, digital computer and control systems, airborne electronic and control systems, electronic instrumentation and test equipment and basic aeronautical and electronic research'*
>
> *(Anon 1956: p. 36S).*

The development of Atlas was managed and coordinated by the Air Force's Western Development Division led by General Schriever and Ramo-Wooldridge (Johnson 1997; Hughes 1998; Morris 2013). Whereas project management, with concerns for military objectives and contractual control, was undertaken by the Western Development Division, Ramo-Wooldridge was responsible for systems engineering and systems integration, with a focus on technical advice and technical direction (Hughes 1998, pp. 116–124). Thus, in this and other similar projects, project managers presided over the technical work performed by the systems integrator as part of a wider remit including concerns such as budgets, deadlines, and scope, but there was a separation of responsibility for technical coordination.

3.3.2 Systems Integration and Organisation Theory: Collaboration Through Coordination and Cooperation

As systems integration was becoming a practical concern on projects, systems became a theoretical concern, with scholars such as Von Bertalanffy (1968) highlighting the non-linear interactions and behaviours that arise. The integration of systems requires collaboration to address such emergent properties and integrate systems. We frame this collaboration as achieved through coordination as the ability to collaborate, and cooperation as the willingness to do so (Tee et al. 2019). Parallel to practical concerns and experiments in how to achieve that, early research attention was given to two these aspects of collaboration: coordination and cooperation.

The coordination of physical components has been most extensively examined through work on the architecture of complex systems (Simon 1981), interdependence (Thompson 1967), and information processing (Galbraith 1973; Galbraith 1977). Simon (1981) highlights the efficiency of architectures that enable 'near decomposability' of complex systems (where intra-component linkages are greater than inter-component links) and the efficiency of the hierarchy as a form of organisation for decision-making. This work has a strong focus on rational (or boundedly rational) decisions to achieve technical coordination and has influenced the development of a trajectory of work on modularity (Baldwin and Clark 2000; Schilling 2000; Baldwin and Clark 2006; Tee 2019), and its role in project complexity (Hellström and Wikström 2005; Tee et al. 2019).

Cooperation is explored in Sayles and Chandler's (1971) work on the Apollo Program (also discussed in Söderlund 2012). Building on Lawrence and Lorsch (1967), they identify systems integration as a key task, describing the roles required by intermediaries to translate across specialisations (Sayles and Chandler 1971, p. 236). They draw attention to how such integration challenges arise across the boundaries of the project, as the evolving context of the project requires cooperation across a wide range of different organisations, including government, firms, and non-government organisations. In a dynamically changing environment, the systems integrator cannot assume that all the ill-defined interfaces – or boundaries between sub-systems – have been identified up front. Using persuasion and bargaining – rather than confrontation and control – to 'induce cooperation', the systems integrator must engage in a continuous interaction with organisations responsible for subsystem development to make any technical problems or ill-defined interfaces 'visible' and work collaboratively to develop solutions to overcome them (Sayles and Chandler 1971, p. 263).

Work on systems thinking complemented by foundational work on the temporal nature of the process of systems integration, including how cooperation is achieved given the diverse incentives and organisational cultures that arise across the boundaries in inter-organisational projects (Sayles and Chandler 1971). Klein and Meckling (1958) recognised in their study of weapon systems projects that knowledge of the component parts of a system may develop unevenly and small adjustments have to be made to integrate components that are 'out of phase' with others into a whole (discussed in Brady et al. 2012). Technology trajectories and changing technological knowledge began to be explored by scholars interested in the relation between innovation and systems integration (Sapolsky 2003). Revisiting the Polaris Program (Sapolsky 1972), Sapolsky (2003) described systems integration as requiring capabilities that are *synchronic*, to coordinate across known components and disciplines, and that are *dychronic*, to coordinate several trajectories of technology development. By addressing both systems integration and technological innovation, such work provides a starting point for unpacking systems integration capabilities at firm and industry levels (Rothwell 1994; Dosi et al. 2003) and starts to extend understanding beyond a static understanding of the physical components and knowledge to be integrated towards a more processual understanding.

Coordination and collaboration are, thus, the two organisational problems underlying the fundamental integration challenge in projects (Söderlund 2011). The problem of coordination arises due to task complexity and the problem of cooperation arises as a result of conflicting goals and opportunistic behaviour (Söderlund 2011). Collaboration is achieved through addressing these problems of coordination and cooperation and is required to deliver integrated systems.

3.3.3 Systems Integration and Construction

While Morris (2013) frames 'systems integration' and the associated practices of 'interface management' as a natural locus for project management, the exact nature of the relationship between systems integration, project management, interfaces, and the organisations in project-based industries has been understood in different ways.

Project management and systems integration responsibilities were co-located but performed separately by two distinct organisations on the Atlas Project (Davies 2017, pp. 46–48). This separation has been seen as less successful on subsequent projects, which require cooperation across a wide range of organisations (Sayles and Chandler 1971). Hughes (1998) describes how the diffusion of system engineering concepts from seminal twentieth-century aerospace and defence projects, such as Atlas, which had a single goal and limited interactions with external environment, exasperated the problems on large construction projects such as the Boston Big Dig, which was a more organisationally complex open system including multiple and diverse stakeholders than envisioned in the processes applied.

Hence, our understanding of construction projects as interventions, evolving, and heterogeneous in nature suggests the need for a different model of systems integration in construction. There is a need for systems integration activities not only to focus on coordination, but also to focus on the cooperation that motivates such collaboration, and there is a need to focus not only upstream, working with the supply chain to integrate sub-systems and but also *downstream* with operators and end users as systems are delivered into contexts of use. Systems integration in manufacturing has been understood as involving work in-house, which then forms the basis for systems integration at the interfaces and integrating the work of supply chain members. Smyth (2014) frames systems integration as starting in the construction firm to overcome the silos between functions and departments and then used in capabilities at the project level. Morris instead focuses on the project and argues that 'the project manager is 'the single point of integrative accountability' in achieving the outcome desired by the sponsor, and to the extent possible, the other stakeholders' (Morris 2013, p. 284).

We argue that clients need to play a significant role in systems integration on extremely complex projects as they retain accountability, however – as noted above – scholars have made a variety of alternative claims regarding the locus of responsibility for systems integration. Winch (1998), for example, argues that the systems integrator role is shared between the principal architect/engineer and the principal contractor in construction. However, such firms operate as service providers and lack focus on outcomes due to the dominance of low price competitive tendering and high risk levels (Smyth 2018). On very large megaprojects and client programmes, the client needs to take accountability and responsibility or needs to work with an independent systems integrator to coordinate at a higher level than any one supplier firm operates.

3.4 Systems Integration in Construction and Contexts of Use

So far, we have outlined the challenges of system integration in construction, as (i) projects are interventions, (ii) they are evolving phenomena, and (iii) the sector is heterogeneous. We have rehearsed the origins of systems integration and the role of cooperation as well as coordination in addressing complexity and uncertainty. In this section, we expand our argument in relation to systems integration in construction, highlighting how construction is involved with open systems that include outcomes and the context of use.

It is important to recognise that in construction, project complexity and uncertainty arise not only in the organisation of delivery and delivered system, but also in the contexts of use (Brady and Davies 2014). In these contexts of use, there is a need to better understand delivered systems as interventions into wider systems of systems (Whyte et al. 2019). This is vital if construction projects are to deliver safety-critical systems, net zero pollution and higher quality through offsite manufacture and assembly, and digital platform approaches.

Project deliverables have far-reaching interconnections and effects. The purpose of a project is to solve a business problem, social goal, or policy objective (Smyth 2018), although the integration into the wider system is often not measured as part of the business case for the project. An exception is the legacy objective of the London 2012 Olympics, which drew attention beyond the delivery of international games to the regeneration of the social, economic, and ecological fabric of East London. Early scholars, such as Sayles and Chandler (1971), and later Hughes (1998), pointed to this need to address projects as organisationally complex open systems, but this is not fully realised. We expand the discussion of systems integration processes *within* projects (intrinsic) to address complexity and uncertainty *across* project boundaries (extrinsic) with interfaces and interdependencies across the systems of systems associated with the delivered system. While Brady and Davies (2014) use Morris and Hough's (1987) concept of extended systems, to identify and address the broader social and political context and communities affected by projects, we develop a more encompassing view that includes the wider technological and ecological contexts. Thus, as shown in Figure 3.2, complexity and uncertainty across project boundaries may have socio-political, technological, and ecological aspects.

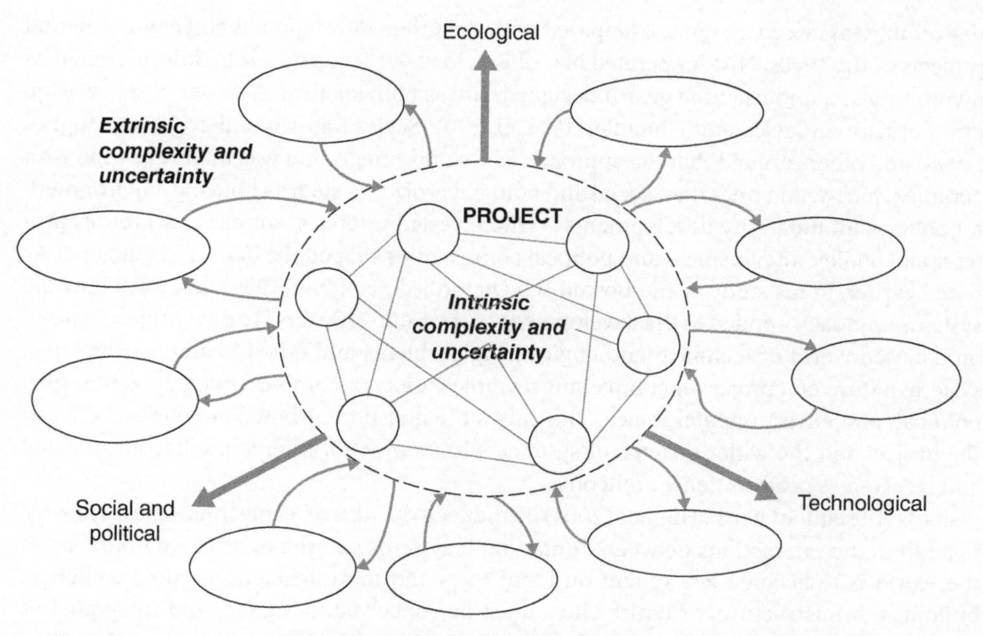

Figure 3.2 Interfaces and interdependencies across the system of systems associated with the delivered system, building based on Grafius et al. (2017).

Recognising the extrinsic as well as intrinsic complexity and uncertainty along these dimensions is vital to achieving safety-critical systems and net zero pollution and to using digital manufacturing.

- *Safety-critical systems:* In many instances, including retrofit, the need for oversight and governance of the whole system with clarity of roles and responsibilities, and a 'golden thread of information' (Hackitt 2018), also requires an understanding of the porous boundary between the project and its interfaces and interdependencies with wider systems of systems. Here, Winch's (2009) information processing perspective may be useful to track how information is shared to enable the delivery of safety-critical systems.
- *Net zero pollution:* Across the interfaces of the project, there are new needs to understand the environmental impact of processes in the supply chain, and also how the construction project impacts natural ecologies, around the site in delivery, in operations, and across the global supply chains that deliver materials, as well as how the changing context of the environment impacts on the way that projects are managed.
- *Modularity:* The portfolio approach required to gain value through the use of digital product platforms suggests a need to consider innovation across a set of projects, and this requires the consideration of learning across projects. Portfolio management and project portfolio management are relatively poorly developed in construction (Smyth and Wu 2021) although there are pressures for firms to modularise, and firms are beginning to use platform strategies (Mosca et al. 2020; Jones et al. 2021; Zhou et al. 2021).

In summary, across the project boundaries, there are interfaces and interdependencies that need to be appreciated by the project team and that involve significant complexity and uncertainty. A criticism of early work on systems integration is the tendency to see project deliverables as closed systems. Compared with the urban development and environmental projects of the 1960s, NASA operated in a 'closed loop – it set its own schedule, designed its own hardware, and used the gear it designed. It was both sponsor and user. Space was no one's territory' (Sayles and Chandler 1971, p. 320). Sayles and Chandler (1971), Hughes (1998) and others argued that the approach to systems integration was unable to cope with complex, messy, and uncertain social and political projects – such as housing, environmental policy, and inner-city developments – where vested interests, community groups, professional bodies, and business and political complexities shaped the delivered outcome. As noted earlier, in his study of the Boston Big Dig project, Hughes (1998) identifies how the systems approach applied to the development of the Central Artery/Tunnel project focused on a closed world of technical and engineering problems and failed to address the 'open system' nature of a project encompassing a whole variety of complex and emergent social, political, and environmental issues. This raises the question of how boundaries, between the project and the wider systems of systems within which it intervenes, become created and resolved as projects deliver outcomes.

In his subsequent work, Hughes (2004) introduced the idea of 'ecotechnological systems' to address the interactions between human-built systems and the natural environment. If the world is to become less reliant on fossil fuels and more resilient to climate change, buildings, infrastructure, or entire cities must be conceived, designed, and integrated as ecotechnological systems to account for natural factors, such as local climate, ecology, and resources. For example, the British architectural firm Fosters + Partners have designed a new city, called Masdar near Abu Dhabi, as an integrated system of interacting components reliant on solar energy, renewable resources, with a zero-carbon and zero-waste ecology (Stern 2009, p. 141).

Today, many projects, for example infrastructure or information systems projects, also have technical deliverables that are themselves embedded into wider technological systems. They are open systems, in which projects are interventions (Whyte et al. 2019). Processes of systems integration need to address the technological complexity that arises as project deliverables increasingly become part of wider cyber-physical systems, with physical deliverables becoming controlled by and interdependent with digital deliverables. Examples include the Crossrail project, in which existing railways, stations, and signalling systems were upgraded and linked to new infrastructure, fundamentally changing the operation of London's entire transport network as well as creating a new railway.

It is surprising that little contemporary research has responded to the call to treat projects as open rather than narrowly focused closed systems by exploring the extended complex and dynamically changing interdependencies between the delivered system and wider technological, ecological, and socio-political components. We suggest that systems integration in construction project organising requires a view of projects as an intervention into systems of systems to address a wider issue or problem.

Thus, as we suggested above, projects vary in the number of interfaces to be managed and degree of project complexity (Shenhar and Dvir 2007; Brady and Davies 2014), but also in the extent to which these interfaces and the complexity are intrinsic or extrinsic to the project. A small repair to a water or transport infrastructure, for example, may have little internal systems complexity but entail significant external interfaces and interdependencies. In such projects, beyond the interfaces and interdependencies within a system or a system of systems, delivered by a project, there is a need to consider how the connections across project boundaries, and with wider built and natural systems, become important to realise value from projects.

3.5 Conclusions

We frame systems integration as an open-ended challenge for construction project organising. We argue that an outcome-focused approach to systems integration is a particular challenge as projects intervene into wider systems of systems and are evolving in nature, and as the sector is heterogeneous. Having rehearsed the origins of systems integration and the role of cooperation as well as coordination in addressing complexity and uncertainty, we draw particular attention to how construction is involved with open systems with complexity and uncertainty across project boundaries having socio-political, technological, and ecological aspects.

Construction brings the challenges of moving from a closed aerospace model for the cold war to an open system more appropriate to complex social, political, and ecological built environments with multiple stakeholders. Our model of systems integration, thus, shifts the attention of project sponsors, project leaders, and project governance from contracts, and managing contracts, to recognising the intrinsic and extrinsic complexity and uncertainty arising in construction projects and managing the delivery of systems and outcomes in complex, messy and uncertain contexts. As many sources of extrinsic complexity and uncertainty are not under the project's control, this requires substantial focus on cooperation, as well as coordination, to achieve desired outcomes.

Systems integration needs to be given priority in delivery and be supported by rather than managed through contracts, budgets, and schedules. While in very simple, repetitive projects, it may be possible to devolve responsibility for systems integration to a consultant or to the main contractor, in highly complex projects, we see the client as accountable for systems integration. The client cannot fully outsource this responsibility and must retain the oversight of the overall systems and interfaces. Research should build on the idea that projects are interventions to account for the open, interacting, and messy interactions between the system delivered and the wider technological, ecological, and social systems that encompass them. So, construction raises wider contextual challenges for systems integration and how this is addressed by clients or contractors in projects, programmes, and portfolios.

These insights have practical import and suggest new areas for research. There are areas for further research to give attention to those boundaries in systems of systems: how they are created and overcome in project organising. We contend that work on systems integration can address areas of contemporary concern such as the delivery of safety-critical systems, net zero pollution and construction transformation to use offsite manufacture and assembly, and digital platform approaches, and that each of these areas require a more expansive understanding of open systems. There is scope for empirical studies to examine how practitioners can best apply insights on systems integration to each of these challenges, and in the emerging tensions across them, where for example offsite and digital approaches require an early design fix and reduce the ability for late design changes. The question of systems integration is of significant importance in construction with increasingly open-ended and complex interdependencies across project boundaries.

References

Anon (1956). The Ramo Wooldridge Corporation. *Journal of Jet Propulsion* 26 (5): p37S.

Baldwin, C.Y. and Clark, K.B. (2000). *Design Rules: The Power of Modularity*. Cambridge, MA: MIT Press.

Baldwin, C.Y. and Clark, K.B. (2006). Modularity in the design of complex engineering systems. In: *Complex Engineered Systems* (ed. D. Braha, A.A. Minai and Y. Bar-Yam), 175–205. Springer.

Biersteker, E., Koppenjan, J., and van Marrewijk, A. (2021). Translating the invisible: governing underground utilities in the Amsterdam Airport Schiphol terminal project. *International Journal of Project Management* 39 (6): 581–593.

Blockley, D. and Godfrey, P. (2017). *Doing it Differently: Systems for Rethinking Infrastructure*. ICE Publishing.

Brady, T. and Davies, A. (2014). Managing structural and dynamic complexity: a tale of two projects. *Project Management Journal* 45 (4): 21–38.

Brady, T., Davies, A., and Nightingale, P. (2012). Dealing with uncertainty in complex projects: revisiting Klein and Meckling. *Special Issue: Classics of Project Management, International Journal of Managing Projects in Business* 5 (4): 661–679.

Davies, A. (2017). *Projects: A Very Short Introduction*. Oxford University Press.

De Graaf, R.S., Vromen, R.M., and Boes, J. (2017). Applying systems engineering in the civil engineering industry: an analysis of systems engineering projects of a Dutch water board. *Civil Engineering and Environmental Systems* 34 (2): 144–161.

De Weck, O.L., Roos, D., and Magee, C.L. (2011). *Engineering Systems: Meeting Human Needs in a Complex Technological World*. MIT Press.

Dosi, G., Hobday, M., Marengo, L., and Prencipe, A. (2003). The economics of systems integration: towards an evolutionary interpretation. In: *The Business of Systems Integration* (ed. A. Prencipe, A. Davies and M. Hobday), 95–113. Oxford, UK: Oxford University Press.

Galbraith, J.R. (1973). *Designing Complex Organization*. Reading, MA: Addison-Wesley.

Galbraith, J.R. (1977). *Organization Design*. Reading, MA: Addison-Wesley.

Gholz, E., James, A.D., and Speller, T.H. (2018). The second face of systems integration: an empirical analysis of supply chains to complex product systems. *Research Policy* 47 (8): 1478–1494.

Gil, N.A. and Fu, Y. (2022). Megaproject performance, value creation and value distribution: an organizational governance perspective. *Academy of Management Discoveries* 8 (2): 224–251.

Grafius, D., Kim, H. and Whyte, J. (2017). Ecological Interdependencies of Infrastructure Projects. *International Symposium on Next Generation Infrastructure* (11 to 14 September). London, UK.

Hackitt, J. (2018). Building a Safer Future: Independent Review of Building Regulations and Fire Safety: Final Report, CM9606, London.

Hall, J.W., Henriques, J.J., Hickford, A.J., and Nicholls, R.J. (2013). Systems-of-systems analysis of national infrastructure. *Proceedings of the Institution of Civil Engineers-Engineering Sustainability* 166 (5): 249–257.

Hellström, M. and Wikström, K. (2005). Project business concepts based on modularity-improved manoeuvrability through unstable structures. *International Journal of Project Management* 23 (5): 392–397.

Hobday, M., Davies, A., and Prencipe, A. (2005). Systems integration: a core capability of the modern corporation. *Industrial and Corporate Change* 14: 1109–1143.

Hughes, T.P. (1998). *Rescuing Prometheus*. New York: Pantheon Books.

Hughes, T.P. (2004). *Human-Built World: How to Think about Technology and Culture*. University of Chicago Press.

ICE (2020). *A Systems Approach to Infrastructure Delivery: A Review of How Systems Thinking Can Be Used to Improve the Delivery of Complex Infrastructure Projects*. London, UK: Institution of Civil Engineers.

INCOSE (2014). *SE Vision 2025: A World in Motion*. International Council on Systems Engineering.

IPA (2021). Systems Integration Module. Project Routemap. London, Infrastructure and Projects Authority.

Johnson, S.B. (1997). Three approaches to big technology: operations research, systems engineering, and project management. *Technology and Culture* 38 (4): 891–919.

Johnson, S.B. (2003). Systems integration and the social solution of technical problems in complex systems. In: *The Business of Systems Integration* (ed. A. Prencipe, A. Davies and M. Hobday). Oxford: Oxford University Press.

Johnson, S.B. (2013). Technical and institutional factors in the emergence of project management. *International Journal of Project Management* 31 (5): 670–681.

Jones, K., Mosca, L., Whyte, J. et al. (2021). Addressing specialization and fragmentation: product platform development in construction consultancy firms. *Construction Management and Economics* 1–16.

Klein, B. and Meckling, W. (1958). Application of operations research to development decisions. *Operations Research* 6 (3): 352–363.

Kouassi, A. (2015). Systems integration. INCOSE infrastructure working Group.

Lawrence, P.R. and Lorsch, J.W. (1967). Differentiation and integration in complex organizations. *Administrative Science Quarterly* 1–47.

Mahnken, T.G. (2008). *Technology and the American Way of War Since 1945*. Columbia University Press.

Morris, P. (2013). *Reconstructing Project Management*. Chichester: Wiley.

Morris, P.W.G. and Hough, G.H. (1987). *The Anatomy of Major Projects – A Study of the Reality of Project Management*. Wiley.

Mosca, L., Jones, K., Davies, A. et al. (2020). Platform Thinking for Construction, Transforming Construction Network Plus, Digest Series, No. 2.

Rothwell, R. (1994). Towards the fifth-generation innovation process. *International Marketing Review* 11 (1): 7–31.

Rutten, M., Dorée, A., and Halman, J. (2009). Innovation and interorganizational cooperation: a synthesis of literature. *Construction Innovation* 9 (3): 285–297.

Sapolsky, H.M. (1972). *The Polaris System Development: Bureaucratic and Programmatic Success in Government*. Cambridge, MA: Harvard University Press.

Sapolsky, H.M. (2003). Inventing systems integration. In: *The Business of Systems Integration* (ed. A. Prencipe, A. Davies and M. Hobday), 15–34. Oxford: Oxford University Press.

Sayles, L. and Chandler, M.K. (1971). *Managing Large Systems*. New York, NY: The Free Press.

Schilling, M.A. (2000). Toward a general modular systems theory and its application to interfirm product modularity. *Academy of Management Review* 25 (2): 312–334.

Selznick, P. (1953). *TVA and the Grass Roots: A Study in the Sociology of Formal Organization*. University of California Press.

Shenhar, A.J. and Dvir, D. (2007). *Reinventing Project Management: The Diamond Approach to Successful Growth and Innovation*. Harvard Business Review Press.

Simon, H. (1981). *The Sciences of the Artificial*. Cambridge: MIT Press.

Smyth, H. (2014). *Relationship Management and the Management of Projects*. Routledge.

Smyth, H. (2018). Projects as creators of the preconditions for standardized and routinized operations in use. *International Journal of Project Management* 36 (8): 1082–1095.

Smyth, H. and Wu, Y. (2021). *Project Portfolio Management: Practice-Based Study of Strategic Project Portfolio Management in the UK Construction Sector*. London: Bartlett School of Sustainable Construction.

Söderlund, J. (2011). Theoretical foundations of project management. In: *The Oxford Handbook of Project Management* (ed. P.W. Morris, J.K. Pinto and J. Söderlund). Oxford: Oxford University Press.

Söderlund, J. (2012). Project management, interdependencies, and time: insights from managing large systems by Sayles and Chandler. *International Journal of Managing Projects in Business* 5 (4): 617–633.

Staal-Ong, P.L., Kremers, T., Karlsson, P.-O., and Baker, S. (2016). *10 Years of Managing Large Infrastructure Projects in Europe: Lessons Learnt and Challenges Ahead*. Netherlands: NETLIPSE.

Stern, N. (2009). *A Blueprint for a Safer Planet: How to Manage Climate Change and Create a New Era of Progress and Prosperity*. Random House.

Sydow, J. and Braun, T. (2018). Projects as temporary organizations: an agenda for further theorizing the interorganizational dimension. *International Journal of Project Management* 36 (1): 4–11.

Tee, R. (2019). Benefiting from modularity within and across firm boundaries. *Industrial and Corporate Change* 1–18.

Tee, R., Davies, A., and Whyte, J. (2019). Modular designs and integrating practices: managing collaboration through coordination and cooperation. *Research Policy* 48 (1): 51–61.

Thompson, J.D. (1967). *Organizations in Action: Social Science Bases of Administrative Theory*. New York: McGraw-Hill.

Von Bertalanffy, L. (1968). *General System Theory*. New York: George Braziller.

Whyte, J. and Davies, A. (2021). Reframing systems integration: a process perspective on projects. *Project Management Journal* 52 (3): 237–249.

Whyte, J., Fitzgerald, J., Mayfield, M. et al. (2019). Projects as Interventions in Infrastructure Systems-of-Systems. *INCOSE International Symposium* (July 20-25), Orlando, Florida.

Whyte, J., Mijic, A., Myers, R.J. et al. (2020). A research agenda on systems approaches to infrastructure. *Civil Engineering and Environmental Systems* 37 (4): 214–233.

Whyte, J., Davies, A., and Sexton, C. (2022). Systems Integration in Infrastructure Projects: Seven Lessons from Crossrail. *Management, Procurement and Law* 175 (3): 103–109.

Winch, G. (1998). Zephyrs of creative destruction: understanding the management of innovation in construction. *Building Research & Information* 26 (5): 268–279.

Winch, G.M. (2009). *Managing Construction Projects*. Wiley.

Zhou, A., Mosca, L., Hall, D. and Whyte, J. (2021). Open Platform and Value Appropriation: Cases from Digitally-Enabled Product Platforms IPDMC (Innovation and Product Development Management Conference). June 6-8, Milan, Italy.

4

Organising Project Finance

D'Maris Coffman and John Kelsey

Bartlett School of Sustainable Construction, Faculty of the Built Environment, University College London

4.1 Introduction

Project finance refers to the practice of financing capital projects via their associated projected cash flows and collateralisation of the underlying asset without recourse to the balance sheets of their sponsors. This mode of financing, which is different from corporate finance, has historical antecedents in the early modern period and even in antiquity. It came into its own in the fourth quarter of the twentieth century as projects were increasingly recognised as making a significant contribution to economic activity in the financial sector. Modern project finance is, thus, understood as a method of finance, whereby the borrower is generally a standalone corporate entity, often a special purpose vehicle (SPV) and, therefore, a single-project enterprise. This entity raises equity that together with the project loans provides the finance to create project assets, which can be used to generate cash flows.

The assets are, however, usually highly specific without a ready secondary market. Project loans are, therefore, secured on the operating cash flows rather than the assets themselves. Additionally, lender's intervention rights are written into loan agreements to ensure that the project proceeds to plan and that cash flows are generated in the event of default by the borrower – particularly during the asset creation phase of the project. The advent of large-scale securitisation of project finance debt instruments helped to attract significant institutional capital, which had shied away from infrastructure debt because of the risk/return profile. This chapter argues that despite what some economic theory might predict, the manner in which a project is financed imposes formal 'constraining and enabling' conditions at all phases of the built-asset lifecycle (Cardinale 2018). Successful project organisation and delivery depends upon an adequate understanding of these dynamics.

In contemporary UK Private Finance projects that create new infrastructure facilities, there are fundamentally two types of arrangement. In the first, the SPV creates and operates the whole facility and its designated activity such as a prison. In the second case, the SPV creates a facility but operates only the facility alongside the main sponsor organisation, which

Construction Project Organising, First Edition. Edited by Simon Addyman and Hedley Smyth.

operates the designated activity such as a hospital. In both cases, the contractual operating period is normally for a period of 25–35 years before the facility and its ownership is returned to the state. In the former case, the commissioning government department acts as the sponsor and the SPV carries out the full client role. In the latter case, the parent government department and the activity manager (such as a National Health Service [NHS] Trust) acts as the sponsor as well as the client role in briefing, outline design, and operation. The SPV acts as a client in the later design, procurement, and construction phases as well as post-completion facility management.

4.2 Economic and Finance Theory

There are two distinct strands of economic theory, which furnish a basis for arguing that choices in financial structures produce differences in project organisation, project delivery, and the associated welfare outcomes. How far theoretical justifications for project finance survive empirical investigation turns on the assumptions made about the viability of alternatives, whether on-balance corporate financing of built assets or traditional public procurement, and ultimately upon assessments of project value. As illustrated by the Peterborough City Hospital case, the manner in which a project is financed not only constrains and enables project organisation, but, as Cardinale (2018) predicts, also serves to actively orient the organisation of the project.

4.2.1 The Basis of Project Value

Before a project can be financed, it must be evaluated to determine the project value. Compound interest had been understood for hundreds of years, but it was not until the sixteenth century that systematic treatments of present value were published (Trenchant 1561; Stevin 1582). Discounted cash flow came into regular use for evaluating industrial projects in UK coal mining during the period 1772–1810 (Brackenborough et al. 2001). Economists eventually took this up culminating in the work of Fisher (1907, 1930) who provided the general type of formula that is now used evaluating net present value (NPV) and other valuation indicators.

The standard format for evaluating projects is the calculation of NPV represented by

$$\text{NPV} = \sum_{t=0}^{n} \frac{V_t - C_t - K_t}{\left(1+r\right)^t} - K_o$$

where t is the relevant project period, n is the project length, V_t is the period revenue, C_t is the period operating cost, K_t is the end of project remediation and restoration cost, r is the discount rate, and K_0 is the initial capital cost.

4.2.2 Sources of Forecasting Error

Forecasting is required to estimate the project value. Simon (1947) challenged the idea that we are fully rational decision-makers because of our cognitive limitations, which he called 'bounded rationality'. Tversky and Kahneman (1973, 1974) and Kahneman and Tversky (1979) show that people often make irrational judgements about extreme or rare events either because of: (i) falsely overestimating the significance of certain characteristics of the forecast situation, or (ii) by unjustifiably relying on heuristics or past 'rules of thumb', or (iii) by mis-specifying the forecasting problem, or (iv) by misinterpreting cause and effects from regression data, or a combination of these. Moreover, constructed assets fall within a category known as complex products and systems. This itself gives rise to problems in project organisation (Hobday 2000). A physicist has suggested that complexity itself provides problems in forecasting (Grassberger 2012).

As a consequence of this, more complex methods of forecasting have been devised. However, practice shows that these frequently add little value to the forecast but are often used to justify or even 'reassure clients through incomprehensibility' (Green and Armstrong 2015). Those providing forecasts are often not accountable for errors, and project input data may be intentionally misleading in order to ensure that a project goes ahead (Flyvbjerg et al. 2003). It is, therefore, likely that there may be substantial underestimates of capital costs, and this is borne out by Mott MacDonald (2002), Flyvbjerg et al. (2003), and Merrow (2011) for both the public and private sector projects. Recognition of this is important as forecasting errors in demand for infrastructure assets can significantly undermine their business case and ultimately their financial performance.

4.2.3 Why Do Firms Exist?

To understand the economic logic for creating separate project enterprises, it is first necessary to consider why organisations, particularly firms but also including separate project organisations such as SPVs, exist at all. Economists were surprisingly slow to address this issue in depth. Coase (1937) pointed out that all firms effectively suppress the price mechanism because information, search, negotiation, contract, and repeat costs for certain types of market transactions (augmented due to risk, uncertainty, and asymmetry of information) exceed the costs of eliminating or minimising them (transaction costs) through internalisation within a single organisation. This is a particular issue where stability of long-term supply and distribution networks is required. The extent to which the firm as a form of organisation promotes productive efficiency depends on the nature of the economic activity undertaken.

The application of Coase's theories to explain the development of SPVs to deliver infrastructure assets is a natural step, one which project finance can be said to facilitate, despite the transaction costs associated with the establishment of the SPV. Debates about the nature and the boundaries of the firm also naturally lead to questions of maximum efficient scale and scope of a given enterprise. Thus, it is reasonable to suppose that the choice of project finance can affect not only productive efficiency, but also allocative efficiency, and thus attendant welfare outcomes. This form of project organising can give rise to agency costs within a firm and between different parties such as owners and managers.

4.2.4 Managers vs. Owners and the Basis of Firm Value

The problem of managers potentially acting in their own interests to the detriment of the wider ownership of the firm was noted as far back as Smith (1776/1982) and later by Berle and Means (1932). Jensen and Meckling (1976) note adopted solutions such as 'bonding costs' (such as commitment to significant share ownership) incurred by managers to disincentivise their own potential misbehaviour and monitoring costs incurred by shareholders to detect and/or limit managerial misconduct (such as external audit and board committees). In spite of this, 'residual agency costs' may still occur.

Projects involving the creation of large physical assets are likely to generate large cash flows, which offer an opportunity for managerial misconduct. Esty (2003) notes that having a single-project company with a relatively high level of debt acts as a deterrent since a significant portion of the cash is earmarked for debt interest payment, and the lack of other projects makes the remaining cash easier to track. Additionally, the existence of a single-project company acts as a deterrent against strategic opportunistic behaviour by suppliers or joint owners such as the 'hold-up' problem, whereby firms needing to cooperate are wary about moving first in incurring otherwise irrecoverable expense, which may reduce their bargaining power under threat of subsequent non-cooperation by the other party.

A sponsoring firm is likely to already have an existing financial structure, which discourages taking on significant debt. The SPV insulates the new project from the sponsor's balance sheet and helps to discourage 'underinvestment' due to previous debt history (Esty 2003). A large high-risk project can generate large distress costs. Isolation within a single-project company can protect the sponsor from a failed project (and incidentally the project from a failed company). By investing in a particular project, the sponsor does not have to 'bet the farm' on the success of an individual project.

More generally, in corporate finance, it can be said that a company with no debt is underperforming in terms of shareholder returns. Taking on new debt and investment is generally a good sign to the market that the company is being well-run with growth prospects unless it is judged that the new debt levels are unsustainable.

Modigliani and Miller (1958) proposed a theory that the allocation of corporate finance between shareholders and lenders should not affect the value of a firm. This was based on rather restrictive assumptions, which excluded the consideration of agency and bankruptcy costs as well as taxation. Stiglitz (1969, 1973) demonstrated that bankruptcy costs did affect the company value. Esty (2003) does likewise. Agency costs in project finance can be burdensome, which requires the design of solid governance mechanisms and the necessary coordination mechanisms for these to be effective, although governance for project finance does not necessarily translate into solid sound governance for project delivery and value for money in project outcomes.

4.3 Agency Costs and Project Governance

4.3.1 Governance of Large Projects to Minimise Agency Costs

Williamson (1979) explains the governance structure for large-scale projects involving the construction of specialist assets although he did not specifically have project finance in mind. In the detailed design and construction phase and in asset ownership, the SPV assumes the role of the client. At this point, the SPV has very high set-up 'sunk' costs, which are irrecoverable if a project is not carried through to completion. Specialist assets have few if any alternative uses and would pose extreme problems of valuation in transfer to another client – the problem of 'asset specificity'. Traditional corrective market mechanisms for repeat transactions of non-specialist goods do not, therefore, work. Prior to the completion of asset creation, the net realisable value of the SPV is significantly less than either the historic cost to date or the SPV value as a going concern. Two mechanisms are possible to help ensure specific performance, namely third-party arbitration (monitoring) and/or inspection or integration under a unitary governance organisation. Accordingly, lenders insist on the right to 'step in' and manage the project if there is a risk that the asset construction may not be completed and 'step out' when they are satisfied that the project has reached a stable and satisfactory operating stage.

In a well-managed project, the maximum borrowing will be the point at which increasing operating surpluses can just meet loan interest/repayment instalment payments after which cumulative short- and long-term borrowing should decline.

Henisz et al. (2012) show that in large projects, the project coalition membership (clients, finance providers, suppliers, consultants, regulators, etc.) changes over time such that agency costs of prior errors may be locked in and borne at different project stages by other stakeholders through a process of 'displaced agency'. They suggest 'relational contracting' as a governance mechanism. This involves contracts governed by trust, transparency, fairness, and partnership (Colledge 2005). Unfortunately, this is not an approach generally found in UK construction despite repeated efforts by government and industry bodies (e.g. see Latham (1994), Egan (1998), and Wolstenholme (2009)). Additionally, Chang and Ive (2007a) note that asset specificity in construction can lead to a reversal of bargaining power, whereby clients may be forced to concede additional costs to a contractor even where this is not provided for in the contract.

4.3.2 The Whole Life Contract Mechanism as a Means of Minimising Agency Costs

A privately financed concession facility requires a whole life procurement method that delivers single-point responsibility from the sponsor's viewpoint and, thus, mitigates most of the principal–agent problems (Table 4.1). Conversely, it should be noted that whole life procurement does not necessarily require private finance.

Table 4.1 Agent failure linkage to components of project value.

Agent/SPV failure	Effect	NPV consequence
Poor design or build quality	Construction reworks	Increase in K_0
	Reduction in revenue – Greater likelihood of breakdowns	Decrease in V_t
	Greater operating costs	Increase in C_t
Poor procurement and/or cost control	Higher capital costs	Increase in K_0
Design or construction delays	Revenue delays	Decrease in V_1-C_1
	Higher construction cost	Increase in K_0
	Higher capitalised interest	Increase in K_0

(Continued)

Table 4.1 (Continued)

Agent/SPV failure	Effect	NPV consequence
Poor operation and maintenance	Greater likelihood of breakdowns	Decrease in V_t
	Greater operating costs	Increase in C_t
	Greater end-of-life costs	Increase in K_t
Poor marketing and management	Reduction in revenue	Decrease in V_t

Source: authors.

With reference to the project evaluation equation above, whole life procurement imposes a strict financial discipline on the SPV in that any serious errors in project or operations management may have significant negative consequences for NPV. With single-point responsibility, the buck literally stops with the SPV. As a governance mechanism, this is the very opposite of a relational contracting approach.

4.3.3 Long-Term vs. Short-Term Risk

As Sorge (2011) argued, the credit risk profile in project finance differs considerably from that in corporate finance, in that in project finance, the credit risk (as well as associated uncertainties) is highest in both the absolute and relative terms at the inception of a project, but tends to steadily diminish over the life of the project; when corporate finance is used to fund fixed capital investment, the firm's balance sheet is often initially strong enough to secure access to cheap credit, but may deteriorate over time either due to the financing demands of the project itself or due to factors external to it.

Therefore, there is a problem that project risks are generally much greater in the asset creation phase than in the operations phase. The 'bundling' of short- and long-term risks would appear to unnecessarily increase the cost of capital unless the project is refinanced at the start or soon after the commencement of the operations phase.

4.3.4 Project Finance as a Solution to Project Governance

The historical antecedents of modern project finance arose primarily to resolve otherwise intractable problems of project governance. Dating the origins of project finance turns to a degree on what is considered its salient feature. If the emphasis is on non-recourse lending, then Greek and Roman 'sea loans' are the obvious classical antecedents, as the market priced them according to the idiosyncratic risks associated with a specific voyage (Kavaleff 2002). Roman law also provided for 'limited liability' companies for a 'sole-purpose' with the benefit of protecting the personal property of shareholders from exposure to business risks (Kavaleff 2002). If the focus is instead on the use of project finance by state actors, whether republican or monarchical, the usual case cited is the English's crowns use of non-recourse loans from Italian merchants to finance silver mines in Devon and Cornwall (Esty et al. 2014; Kayser 2013; Müllner 2017).

In Renaissance and early modern Britain and the Dutch Republic, the English and Dutch East India companies financed sea voyages through non-recourse loans, which were repaid after the sale of cargo (Esty and Christov 2002). Over time, chartered companies and joint-stock companies developed as competing corporate forms to provide more permanent capital, once it became clear that such arrangements offered vehicles for superior risk diversification, as the upfront costs were high and could not be recouped by individual investors if a given voyage failed (Smith 2018, 2021). The English crown over the seventeenth- and eighteenth-century licenced 'projectors' to raise non-recourse financing for a variety of projects of perceived public benefit, including fen and swamp drainage, fisheries, wreck salvage, and even overseas colonial ventures (Yamamoto 2018). Most financing of public works in the growing trading cities was done by municipal corporations who borrowed on their own balance sheets (Coffman et al. 2022).

From the mid-eighteenth century through to the early twentieth century, the widespread use of project financing was not a feature of western industrialising or industrialised economies, nor particularly of imperial ventures. Project finance as we recognise it today returned in the early-twentieth-century United States, where it provided an attractive vehicle with which to finance oilfield exploration and speculative real estate development (Esty et al. 2014). Until the 1980s, project finance was generally restricted to the oil and gas, mining, and real estate sectors.

Esty and Christov (2002) follow most authors in identifying the passage in 1978 of the Public Utility Regulatory Act in the United States as providing the impetus for the development of the so-called modern project finance. This legislation, designed to encourage the development of renewable and alternative energy, obliged local utility companies to negotiate long-term power-purchase agreements ('take-or-pay' contracts) with qualified power producers, which provided the economic logic for project sponsors to establish power companies (called Independent Power Producers) funded by non-recourse debt collateralised by the power-purchase agreements. Nevertheless, in North America, project finance outside the energy sector was relatively rare until the early twenty-first century, when extensions to the transport sector and social infrastructure received inspiration from the perceived success of the British model.

4.3.5 UK Public–Private Finance Initiative

The UK experience needs to be seen against a context of a public sector counter-revolution from the 1970s onwards following the slowdown in economic growth following the oil crisis in 1973. The combination of warfare and economic depression in the period 1914–1945 had led to greatly expanded state sectors. The key players in the economy became oligopolist large private sector corporations and the state both as employer and large-scale purchaser of goods and services.

This was accompanied by large-scale technical change, which led to a substantial growth in average wages. For capital-intensive industries, this was less of a problem, but for labour-intensive industries and government activities, this became a real problem, which led to public sector costs rising faster than the average rate of inflation and, if unchecked, requiring a greater share of the economy simply to deliver the same services. This led to the 'Fiscal Crisis of the State' (O'Connor 1973). At the same time, there was criticism of the relative efficiency of the public sector, the role of the state as regulator, and that many large organisations (both public and private) had become too big with the cross-subsidisation of inefficient activities.

Parallel with these developments was the rise of monetarist theory and policy in opposition to the prevailing Keynesian orthodoxy. This, in particular, stressed control of the money supply and reining in of public expenditure and borrowing. A more overtly political development was the desire by the Conservative government of 1979–1997 to curb the power of public sector trade unions who many saw as having become too powerful. A final objective was to remove risk from public sector activities subject to overspending and delay.

This led to three major strands of changed policy:

1) Deregulation – in the UK, the regulation of individual industries was reduced, and state bodies were required to purchase using competitive tendering.
2) Privatisation – sections of the public sector and particularly state-owned enterprise were transferred to the private sector on the grounds that they would be better managed as private sector enterprises.
3) Marketisation – large government organisations should be broken up and operated with decentralised business units operating in a quasi-market – the UK NHS being a prime example.

Because of constraints on public sector borrowing, the government also looked for ways in which capital projects could be privately financed. What emerged was the Private Finance Initiative (PFI) and later the Public–Private Partnership (PPP). Essentially, this meant that a private sector vehicle (the SPV) owned the new capital asset and operated it either with or without the accompanying business. On the one hand, this might mean the totally private operation of transport services; on the other hand, the new facility would support and be operated in conjunction with a state organisation such as the UK NHS or the Prison Service. So, the state paid for the services of a constructed facility without actually owning it until the end of the PFI or PPP contract (Ive 2004).

4.4 Methodology

In the following, we present a case study of a UK PFI hospital. The case study illustrates the effects of choices of project finance on project organising. The case is chosen for two reasons: First, it is an example that has its roots in early development of PFI and is connected to the modern day as the UK NHS seeks to implement a new programme of hospital construction (Mohan 2002); second, the case is an example of not just the influence of project financing, but the wider interconnectedness between project organising and government policy.

Our analysis of the case is interpretive, and the objective twofold. First, our focus was on developing a teaching case around the theory we have positioned in the earlier sections of this chapter. Our analysis, therefore, as well as writing up the case, involved an iterative process between the authors over face-to-face meetings and virtual communication to identify themes for how the case can be used across a range of Built Environment postgraduate programmes. Second, in doing so, we looked to weave together these themes with the concepts presented in the theories above so as to provide a conceptual direction for future theorising. We do not see these two objectives as mutually exclusive, and indeed, in the process of writing, we further developed and clarified our understanding of both objectives.

In presenting the case, first, we offer some background context, including the approval of the investment plan within a wider context of a nationwide hospital building programme;

second, we present the business case for the new hospital and the SPV before moving on to present the organisational and governance arrangements; and third, we finish with a discussion on the performance of the new hospital in operation, drawing out three key lessons from this case for construction project organising.

4.5 Case Study: Peterborough City Hospital (National Audit Office 2012, 2013)

4.5.1 Background

Reforms to the UK NHS arose from the impact of 'New Public Management' (1990s–2010s), which aimed to change public services so as to have a greater focus on the customer and operate more like private sector businesses. In the case of the NHS, this required the creation of local business units in the form of NHS Trusts, which are supposed to be financially self-sufficient. Additionally, a quasi-market was introduced in 1994 into the capital projects procurement process such that Trusts were rewarded by actual performance rather than service capacity (Ruane 2016).

Our case is a new hospital project in the Peterborough and Stamford Hospitals NHS Foundation Trust, which was authorised as a Trust in 2004. The Trust was funded by Public Share Capital from the Department of Health on which it was expected to pay a dividend of 3.5%. It had very considerable financial and managerial autonomy but was answerable for financial management to Monitor in this quasi-marketplace. The government established Monitor in 2003 as a new regulatory body to:

- set prices for NHS-funded care in partnership with NHS England,
- enable integrated care,
- safeguard patient choice and prevent anti-competitive behaviour, which is against the interests of patients,
- support commissioners to protect essential health care services for patients if a provider gets into financial difficulties.

Following financial difficulties (caused in part by the new hospital PFI scheme), it was merged with another NHS Trust in 2017 to become the North West Anglia NHS Trust (Hawkes 2014).

In support of its policy for establishing NHS Trusts, in 2002, the UK government established the East of England Strategic Health Authorities (SHAs) as part of a nationwide set of such bodies to:

- develop a coherent strategic framework for services development across all NHS bodies;
- monitor performance of local NHS Trusts and Primary Care Trusts;
- ensure involvement of patients in decision-making for services development.

In 2013, SHAs were abolished and their functions devolved to NHS Property Services and Public Health England.

In 2014, the Public Accounts Committee of the House of Commons issued a report criticising Monitor for failing to carry out its work properly and for failing to recruit enough staff with appropriate expertise. In 2016 it was absorbed into a new body – NHS Improvement.

4.5.2 The New Hospital

The government announced in 2000 that it would provide investment for 100 new hospitals during the following decade, governed through the arrangement briefly described. It was decided that the only way this could be achieved would be through the PFI, in which the hospital is financed, designed, constructed, and maintained for 25–30 years by a private sector entity (SPV) after which it is transferred to the Trust. Under some schemes, the SPV might also supply 'soft' services such as cleaning and catering. In 2001, the Department of Health approved the Strategic Greater Peterborough Health Investment Plan. At the time, in-patient services were delivered from three sites. This was inefficient, and the facilities themselves were outdated. It was recognised that a new modern facility could better deliver the services currently delivered from the three existing sites. It was also recognised that other services could also be delivered from the new site.

The private sector SPV responsible for designing, constructing, and operating the hospital for 35 years is Peterborough (Progress Health) plc financed by:

(a) £50 000 in ordinary shares;
(b) £396 115 000 guaranteed bonds at 5.581% per annum;
(c) £26 273 000 loan notes at 13.5% per annum.

A majority of shares were held by the Brookfield Group (construction). The Brookfield Group was responsible for the actual design and construction. The income is provided through a monthly 'Unitary payment' by the Trust for all services.

The business case for the new hospital took into account that the Trust already had a maintenance backlog of over £200m and that there was great staff dissatisfaction with existing facilities. However, the fundamental flaw in the scheme business case was the assumption that revenue increases from service outputs and savings in operational costs would offset the annual cost of the PFI scheme. Monitor expressed concerns about the affordability of the original hospital scheme. The UK HM Treasury approved the scheme, provided that Monitor's concerns were adequately addressed. In the end, the key body Monitor was effectively overruled by the Department of Health (Gainsbury 2012; Kmietowicz 2012), which in 2007 approved the PFI scheme without the Trust addressing Monitor's concerns. Subsequent value engineering in the design stage of the project resulted in the new hospital's capacity being reduced by 98 beds. The revised scheme went ahead and the hospital was handed over in 2010.

4.5.3 Organisational and Governance Arrangements

Figure 4.1 sets out in graphical form the organisational arrangements that we have described above. What can be seen from Figure 4.1 is that there appears to have been effectively four different governance bodies (two of which no longer exist) that all in some way influenced the financing and cash flow of the project both in its design and build and in operation.

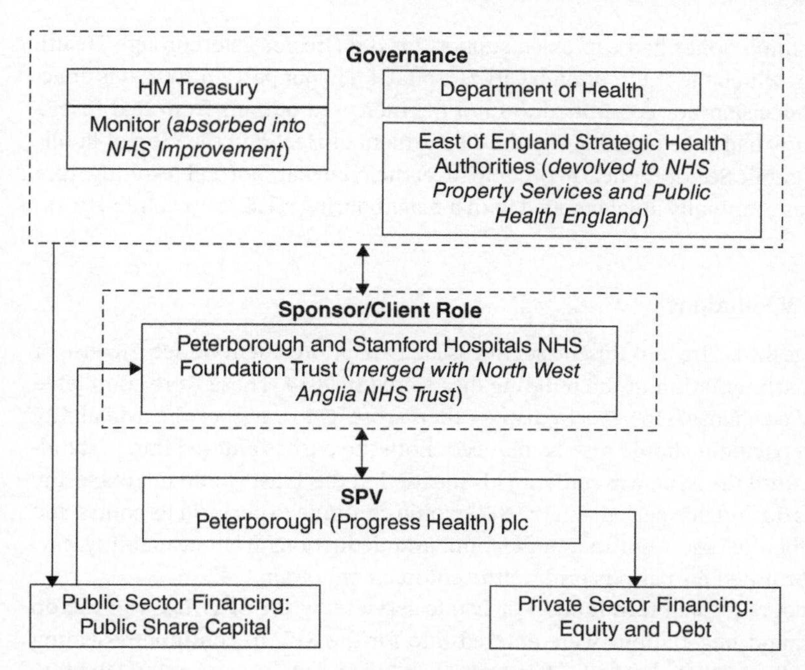

Figure 4.1 Peterborough City Hospital inter-organisational relations. Source: authors.

4.5.4 Operational Performance

The opening of the hospital attracted a considerable number of patients from outside the area. However, the time, cost, and management effort associated with incorporating the new facility into existing arrangements and treating new patients was significantly underestimated and not matched by corresponding income. Therefore, the projected revenue increases and operational cost savings failed to materialise. The Trust soon found itself in serious financial trouble. NHS Peterborough (not to be confused with the Trust), as the main commissioner for the Trust's service outputs, also determines payments to the Trust. It, thus, has a great deal of control over its income.

At the time that the Trust was experiencing financial difficulties with the new hospital, NHS Peterborough was struggling to stay within its Department of Health funding allocation. As a result, it adopted a much stricter regime in performance review and management of contracted services. Accordingly, it reduced payments to the Trust for underachieving against national and locally developed performance indicators. While this was by no means the sole reason for the Trust's difficulties, it made a bad situation worse.

Given the realisation that the Trust was not financially sustainable, the accountants, PwC, were asked to advise on future options. They concluded that the cost of the estate was the major part of the problem, but this included the underutilisation of the estate, which reflects the original downsizing of the scheme. This does ultimately suggest a briefing failure. Additionally, the team concluded that there was no affordable way of refinancing the project suggests that the cost of capital to the Trust is unnecessarily high.

Although the commissioner had expressed support for the Greater Peterborough Health Investment Plan (of which the Peterborough City Hospital is a major part) in 2001, it is unaccountable for that decision nor committed to fund the increased outputs from that investment. It subsequently had to be bailed out by the Department of Health in the form of Public Dividend Capital (Public Sector Equity) in order to meet the financial shortfall resulting from the scheme and was eventually amalgamated with a neighbouring NHS Trust (Limb 2013).

4.5.5 Client–SPV Relations

In 2016, Cambridgeshire Fire and Rescue service issued an enforcement notice because of faults in fire compartmentation in the building discovered in 2014. These were completed in 2018 at the SPV's expense. However, a dispute then arose as to whether the availability part of the unitary payment should also be reduced. Both the parties entered into a 'standstill arrangement' until the issue was settled. This meant that the Trust would not make any further deductions during this period and the SPV would continue to perform its contracted services. Eventually, the case went to adjudication, and deductions from availability payments were not permitted for periods prior to the enforcement order.

In 2015, the Trust requested termination of Estate services by the SPV. The right to do this was disputed, and negotiations were entered into for the SPV to continue providing services. Additionally, in 2016–2017, the Trust made £1.7m in deductions from availability payments. In 2018, the termination request seems to have been withdrawn as a result of settlements on a number of issues. However, with the exception of 2011–2012, deductions from the Unitary payment have been made in every year of operation (Peterborough (Progress Health) plc annual accounts 2010–2021). This suggests that client–SPV relations have been less than smooth during the first decade of operations.

In the following section, we discuss this case against the backdrop of the theory we presented in the first part of this chapter.

4.6 Discussion, Lessons, and Theoretical Challenges

The evidence we have presented in our case study above shows that project financing influences project organising by both enabling and constraining the performance of the project during its development, construction, and in operation. While there is much to learn from an in-depth analysis of this case, for the purposes of this chapter, we discuss three key influencing factors, namely (i) client capability; (ii) risk-bearing capacity; and (iii) governance.

First, the evidence shows that the type of organisation that may be suitable or unsuitable for running a local set of health services is not necessarily an appropriate organisation for the active client role in running a large construction project with project finance. Every large asset construction (capital) project for a client essentially requires the client to undertake a business change project to receive and operate the new facility. One of the key failings in this case was that the Trust very significantly underestimated the staff time and thus costs (for existing and new staff) required to receive and operate the new hospital.

However, the client here faced another set of problems because of scale. The hospital project operating and finance costs (involving a 35-year commitment with costs index-linked to the Retail Price Index) were large in relation to the Trusts annual budget, £41.6m compared with annual turnover of £208m for 2011–2012 (NAO 2012). This left very little financial flexibility for the Trust in terms of future development. It also meant that a small and inexperienced client was facing the combination of a large multinational bank (ABN-AMRO) and contractor (Brookfield) who were the main equity holders and lenders in the SPV. The asymmetry of size itself gave rise to greater bargaining power for the SPV and made principal–agent problems and bargaining power reversal (Chang and Ive 2007b) all the more likely. This manifested itself, for example, in the post-construction disputes regarding fire compartmentation and operating service quality. Although there exist bodies such as NHS Improvement, there appears to be no organisation with the capacity and mission to act either directly or in an advisory capacity supporting the necessary active client role in capital projects with or on behalf of relatively small NHS Trusts ill-equipped for such a role. Such an organisation is recommended for the health sector by Naylor (2017) and more generally by Winch and Leiringer (2016). Procurement through the PFI/PPP route does not transfer all risk from the public sector, and therefore, the client role is essential.

Second, the risk-bearing capacity of the firm (Chang 2015) can be applied to an NHS Trust, and here, the unitary charges to be paid for the hospital dominated the subsequent operational budget of the Trust such that it had no room for financial manoeuvre if other financial performance measures were unsatisfactory. The high unitary charges reflect finance where the short- and long-term risks are 'bundled' – this makes no sense in the operational phase when the short-term risks are no longer present. Since one would expect the SPV to renegotiate loans during the operational phase, it makes no sense for the NHS Trust to continue to pay higher interest charges whether or not the PFI contract is itself terminated (Hellowell 2015).

In 2012, Brookfield Asset Management (group parent company) had turnover of US$18.7bn, non-recourse borrowings of US$41.2bn, other corporate borrowings of US$3.5bn, and equity of US$44.3bn (Brookfield Asset Management 2012). For 2012, ABN-AMRO had an operating income of EUR 7.3bn, assets of EUR 394.4bn, and equity of EUR 15.8bn (ABN-AMRO 2012). The marginal risk posed by the Peterborough City Hospital Project was, therefore, very small in relation to the overall size of these companies. However, the Peterborough and Stamford Hospitals Foundation NHS Trust failed to correctly forecast the increase in excess cost of the estates function (both PFI and non-PFI). This was estimated at £11–£26m, which represented 27–63% of the 2011–2012 Trust deficit of £45m. In the context of this project, what might appear as a small forecasting error for the private sector can be significant for the public client when related to operating rather than capital costs. This also poses a challenge in terms of the relationship between the size of the project and that of the client.

Finally, there was a clear failure of project governance by the Department of Health and the SHA in: (i) not stopping the project when it was only marginally viable with the best assumptions; (ii) not setting out alternative options for the health service delivery problems which the hospital was meant to solve; and (iii) not taking a geographically wider view of the problem (as was indeed eventually done). In the context of a completely artificial government-regulated market, those who have the power to regulate charges (the commissioner, in this case) should be part of the project evaluation team and not be excluded from the project team entirely. Due diligence by the lenders would require them to assure themselves on this point.

Project governance should also ensure that there is a system of a detailed post-project evaluation of costs and benefits. A review of costs is inhibited by the insistence of commercial confidentiality by SPV companies. A review of benefit to cost performance only seems to be carried out when things go wrong, such as in this case. This means that there appears to be little systematic inter-project learning. An early report (House of Commons 2003) was very positive about PFI delivering value for money but warned that it was too early for a full assessment to be made. National Audit Office (2020) states that four out of nine surveyed authorities were dissatisfied with the condition of assets at handover at the end of the PFI contract. This suggests a more general problem, namely that those who took decisions at the front end of PFI projects are unlikely to be held accountable for the asset condition 25–35 years later. It should not come as a surprise that the UK government has abandoned PFI/PPP for new forms infrastructure financing and procurement (HM Treasury 2018).

4.7 Conclusions

Organising project finance is not just about finding the right SPV. Capable organisations responsible for active sponsor, client, and governance roles need to be in place prior to any finance deal. In the case set out in this chapter, there were clearly serious flaws in the business case, the briefing process (both local and regional), full stakeholder engagement, and project evaluation. The form of project finance certainly oriented the manner in which these processes were organised; while this may have transferred some risk away from the client, it did not relieve them (or other health-based stakeholders) of collective responsibility for proper organisation of the project front end, nor did it relieve the Department of Health on behalf of the taxpayer from learning lessons from failure.

Further research would be useful in (i) understanding alternative possibilities for private, non-recourse finance following the demise of PFI; and (ii) designing more coherent forms of cooperative multi-stakeholder organisation in both the project front end and in post-construction transfer to efficient and effective operation of the transferred asset and the associated services in which they are delivered.

The need for new public infrastructure in the twenty-first century is growing at a time when public finances are constrained. For those who commission and govern such projects, their financing and subsequent delivery and operation cannot be taken in isolation. This points towards an urgent need for new forms of project organising at the interface between the client and the SPV.

References

ABN-AMRO Bank N.V. (2012). Annual Report https://assets.ctfassets.net/1u811bvgvthc/2sPVFySyUMWmUETaad1N6A/df93bf2478ef32c9c8d74e28f0513d70/ABN_AMRO_Annual_Report_2012.pdf.

Berle, A.A. and Means, G.C. (1932). *The Modern Corporation and Private Property*. New York: MacMillan.

Brackenborough, S., McLean, T., and Oldroyd, D. (2001). The emergence of discounted cash flow analysis in the Tyneside coal industry 1700–1820. *British Accounting Review* 33 (2): 137–155.

Brookfield Asset Management Inc., (2012). Annual Report https://bam.brookfield.com/sites/brookfield-ir/files/brookfield/bam/annual-reports/2012/bam-financials-annual-2012-f.pdf.

Cardinale, I. (2018). Beyond constraining and enabling: toward new micro-foundations for institutional theory. *Academy of Management Review* 43 (1): 132–155.

Chang, C.Y. (2015). Risk bearing as a new dimension in project governance. *International Journal of Project Management* 33 (6): 1195–1205.

Chang, C.Y. and Ive, G.J. (2007a). The hold-up problem in the management of construction projects: a case study of the Channel Tunnel. *International Journal of Project Management* 25: 394–404.

Chang, C.Y. and Ive, G.J. (2007b). Reversal of bargaining power in construction contracts: meaning, existence and implications. *Construction Management and Economics* 25: 845–855.

Coase, R.H. (1937). The nature of the firm. *Economica* 4 (16): 386–405.

Coffman, D.M., Stephenson, J.Z., and Sussman, N. (2022). Financing the rebuilding of the City of London after the Great Fire of 1666. *Economic History Review* https://doi.org/10.1111/ehr.13136.

Colledge, B. (2005). Relational contracting: creating value beyond the project. *Lean Construction Journal* 2 (1): 30–45.

Egan, J. (1998). *Rethinking Construction: Report of the Construction Task Force*. London: Department of Trade and Industry.

Esty, B.C. (2003). The economic motivations for using project finance. *Harvard Business School* 28: 1–42.

Esty, B.C. and Christov, I.L. (2002). An overview of project finance – 2001 update. *Harvard Business School Supplement* 202-105, April 2002. (Revised May 2003.)

Esty, B.C., Chavich, C., and Sesia, A. (2014). An overview of project finance and infrastructure finance – 2014 Update. *Harvard Business School Background Note* 214-083, June 2014. (Revised July 2014.)

Fisher, I. (1907). *The Rate of Interest*. New York: MacMillan.

Fisher, I. (1930). *The Theory of Interest*. New York: MacMillan.

Flyvbjerg, B., Bruzelius, W., and Rothengatter, N. (2003). *Megaprojects and Risk: An Anatomy of Ambition*. Cambridge: CUP.

Gainsbury, S. (2012). Warnings on PFI scheme went unheeded. *Financial Times* (29 November).

Grassberger, P.J. (2012). Randomness, information and complexity. *arXiv*: 1208.3459

Green, K.C. and Armstrong, J. (2015). Simple versus complex forecasting: the evidence. *Journal of Business Research* 68: 1678–1685.

HM Treasury (2018). *Budget 2018 Policy Paper* (29th October). London: HM Treasury

Hawkes, N. (2014). Peterborough trust seeks partner to take it into the black. *British Medical Journal* 348: g2767.

Hellowell, M. (2015). Borrowing to save: can NHS bodies ease financial pressures by terminating PFI contracts? *British Medical Journal* 351: h4030.

Henisz, W.J., Levitt, R.E., and Scott, W.R. (2012). Toward a unified theory of project governance: economic, sociological and psychological supports for relational contracting. *Engineering Project Organization Journal* 2 (1–2): 37–55.

Hobday, M. (2000). The project-based organisation: an ideal form for managing complex products and systems? *Research Policy* 29: 871–893.

House of Commons, (2003). *The Private Finance Initiative*. Research Paper 03/79. London: House of Commons

Ive, G. (2004). Private finance initiative and the management of projects. In: *The Wiley Guide to Managing Projects* (ed. P.W.G. Morris and J.K. Pinto). Hoboken, NJ: John Wiley and Sons.

Jensen, M.C. and Meckling, W.H. (1976). Theory of the firm: managerial behavior, agency costs and ownership structure. *Journal of Financial Economics* 3: 305–360.

Kahneman, D. and Tversky, A. (1979). Intuitive prediction: biases and corrective procedures. *TIMS Studies in Management Science* 12: 313–327.

Kavaleff, A. (2002). Project finance: contracting and proactive preventive law. *Preventive Law Reporter* 21: 18.

Kayser, D. (2013). Recent research in project finance – a commented bibliography. *Procedia Computer Science* 17: 729–736.

Kmietowicz, Z. (2012). Trust was warned not to sign PFI deal that left it needing a bailout. *British Medical Journal* 344: e4472.

Latham, M. (1994). *Constructing the Team: Final Report of the Government/Industry Review of Procurement and Contractual Arrangements in the UK Construction Industry*. London: HMSO.

Limb, M. (2013). PFI is blamed for financial collapse of Peterborough and Stamford Trust. *British Medical Journal* 346: f3735.

MacDonald, M. (2002). *Review of Large Public Procurement in the UK*. London: HM Treasury.

Merrow, E.W. (2011). *Industrial Megaprojects: Concepts, Strategies and Practices for Success*, 1e. Hoboken NJ: Wiley.

Modigliani, F. and Miller, M. (1958). The cost of capital, corporation finance and the theory of investment. *American Economic Review* 53: 261–297.

Mohan, J. (2002). *Planning, Markets and Hospitals*. London: Routledge.

Müllner, J. (2017). International project finance: review and implications for international finance and international business. *Management Review Quarterly* 67 (2): 97–133.

National Audit Office (2012). *Peterborough and Stamford Hospitals, Report by the Comptroller and Auditor General HC 658 Session 2012–13*. London: NAO.

National Audit Office (2013). *2012–13 Update on the Indicators of Financial Sustainability in the NHS, HC 590 Session 2013–14*. London: NAO.

National Audit Office (2020). *Managing PFI Assets and Services as Contracts End*. London: NAO.

Naylor, R. (2017). *NHS Property and Estates: Why the Estate Matters for Patients*. London: Department of Health.

O'Connor, J. (1973). *The Fiscal Crisis of the State*. New York: St Martin's Press.

Ruane, S. (2016). Market reforms and privatisation in the English National Health Service/ Mercado reforma y privatización en el Sistema Nacional de Salud inglés. *Cuadernos de Relaciones Laborales* 34 (2): 263–291.

Simon, H. (1947). *Administrative Behaviour: A Study of Decision-Making Processes*. Hoboken NJ: Wiley.

Smith, A. (1776/1982). *An Inquiry into the Nature and Causes of the Wealth of Nations*. Penguin: Harmondsworth.

Smith, E. (2018). The global interests of London's commercial community, 1599–1625: investment in the East India company. *The Economic History Review* 71 (4): 1118–1146.

Smith, E. (2021). *Merchants: The Community that Shaped England's Trade and Empire, 1550–1650*. New Haven, CT: Yale University Press.

Sorge, M. (2011). The nature of credit risk in project finance. *BIS Quarterly Review* December.

Stevin, S., (1582). Tafalen van Interest, Antwerp, Christoffel Plantijn.

Stiglitz, J. (1969). A re-examination of the Modigliani-Miller theorem. *The American Economic Review* 59: 784–793.

Stiglitz, J. (1973). On the irrelevance of corporate financial policy. *The American Economic Review* 64 (6): 851–866.

Trenchant, J. (1561). L'Arithmetique de Jean Trenchant departies en trois livres. Lyon, Jove.

Tversky, A. and Kahneman, D. (1973). Availability: a heuristic for judging frequency and probability. *Cognitive Psychology* 5 (2): 207–232.

Tversky, A. and Kahneman, D. (1974). Judgment under uncertainty: heuristics and biases. *Science* 184: 1124–1131.

Williamson, O.E. (1979). Transaction cost economics: the governance of contractual relations. *The Journal of Law and Economics* 22 (2): 233–261.

Winch, G.M. and Loiringer, R. (2016). Owner project capabilities for infrastructure development: a review and development of the "strong owner" concept. *International Journal of Project Management* 34 (2): 271–281.

Wolstenholme, A. (2009). *Never Waste a Good Crisis: A Review of Progress since 'Rethinking Construction and Thoughts for our Future*. London: Constructing Excellence.

Yamamoto, K. (2018). *Taming Capitalism before its Triumph: Public Service, Distrust, and 'Projecting' in Early Modern England*. Oxford: Oxford University Press.

5

Organising for Digital Transformation: Ecosystems, Platforms, and Future States

Bethan Morgan[1], *Eleni Papadonikolaki*[2] *and Tim Jaques*[3]

[1] *Director and Co-Founder at Digital Outlook, London, UK*
[2] *University College London, UK*
[3] *Teaming Worldwide, New York, USA*

5.1 Introduction

While digital change has been experienced in the built environment since the 1950s (Gann 2000) it is now accelerating rapidly. The proliferation of interdependent technologies effecting the products and production of the built environment has greatly increased. The systematic combination of these technologies forms the basis of the cyber-physical systems that underpin its digital transformation. Significant changes are evident in the demands made of our built environment: for example, a recent report from the World Bank estimates that by 2050, more than 70% of the global population will live in cities: the built environment that accommodates this trend will generate some 80% of global GDP. The role of digital technologies is to enable the realisation of such changes. The widespread changes implicit in digital transformation (Papadonikolaki and Morgan 2021) demand that parallel contextual social factors and processes are put in place to enable the use of this technology.

This chapter asks then, in the face of such profound and ongoing change what next for the Architecture, Engineering, and Construction (AEC) industry? If digital technologies are key enablers in addressing these challenges, how does the AEC industry realise the promise of digital transformation and overcome its inherent challenges? How can it organise for digital transformation?

We turn here to the construct of ecosystems and platforms as a promising avenue to address this question and generate insights into how to organise for digital transformation. New business ecosystems and meta-organisations, such as platforms, are transforming industries widely and are underpinned by digital innovations. In the built environment, while organising for digital has been studied in firms (Morgan 2019) and supply chains (Papadonikolaki and Wamelink 2017), the need for a greater understanding of the role of ecosystems and platforms has been more recently noted in scholarly research (Glass et al. 2020).

Construction Project Organising, First Edition. Edited by Simon Addyman and Hedley Smyth.

We focus on organising, drawing attention to actions as opposed to objects – to verbs rather than nouns (Weick 1979). To mobilise this focus, we adopt a sociomaterial theoretical perspective, privileging neither human actors nor technologies (Orlikowski 2007). Sociomaterial studies are underpinned by developing a clear understanding of actors' use of technology and, thus, require a focus on evolving practices. They draw on technological fields, which are aggregations of all organisations that influence the development, use, regulation, or exploitation of a technology, and focus attention on the reciprocity between practice and wider structures (Whyte 2010). In turn, this perspective brings attention to the human processes and standards demanded by digital transformation. Such contextual changes are often neglected, as is evident historically in the built environment (Morgan 2019).

Sociomateriality is a very suitable approach to take for studies of ecosystems and platforms (Gawer and Cusumano 2014), which are key to understanding how value is created and captured (Vargo and Lusch 2004). In ecosystems, platform leaders and complementors compete for governance. In construction, a highly project-based sector, the combination of the project and the platform business model, creates new power dynamics that reshape the role of digital innovations.

In this chapter, we discuss related organising mechanisms and propose a methodology for generating them in the built environment (shown in Figure 5.1). We first discuss modularity and the resulting rise of platforms. Business ecosystems are then presented, using empirical examples from a range of industries to illustrate how they evolve and operate. These principles are then discussed by contrasting the approaches taken by 'digital natives', to those incumbent firms that develop ecosystems through evolutionary processes. We then bring these principles together to explore how a future state thinking could be mobilised to create sustainable ecosystems for a digitally transformed built environment.

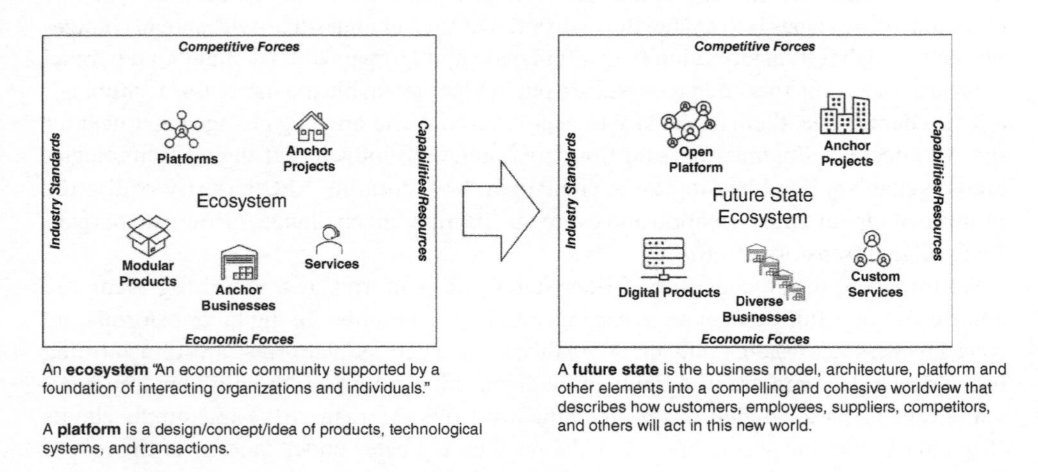

Figure 5.1 Overview of organising mechanisms and future states.

5.2 Modularity, Platforms, and Business Ecosystems

5.2.1 Modularity and Industry Change

At the birth of many industries, companies were integrated, developing all or almost all components themselves, for example at Ford or Toyota. As industries evolved, specialised companies emerged to develop certain components of the larger product, e.g. windshields in cars. This shift is based on modularity principles. Modularity is an approach to addressing increasing complexity and a strategy for organising complex products and processes efficiently (Baldwin and Clark 1999). Modularity is a ubiquitous concept that has proved useful in a large number of fields that deal with complex systems, for example, brain science and psychology, to robotics, artificial intelligence, and industrial engineering (Baldwin and Clark 1999).

A module is a unit whose structural elements are powerfully connected to an overall system and relatively weakly connected to other units (Baldwin and Clark 1999); thus, attention in the built environment shifts from the traditional split between design and construction and operation to seeing the built environment as produced by a complex business ecosystem (this is explored more in Section 5.2.4). A modular system is composed of units (or modules) designed independently but still functioning as a whole. A complex system can be managed by dividing it up into smaller modules separately. Modularity relates to abstraction, information hiding, and interface management. The abstraction hides the complexity of the element; the interface indicates how the element interacts with the larger system (Baldwin and Clark 1999).

Designers achieve modularity by precisely partitioning information into visible design rules and hidden design parameters. The visible design rules (visible information) are used in making decisions that affect subsequent design decisions. Visible design rules fall into three categories: (i) overall system architecture; (ii) interfaces that describe in detail how the modules will interact, including how they will fit together, connect, and communicate; and (iii) standards for testing a module's conformity to the design rules and for measuring module performance. The hidden design parameters pertain only to the relation between the module and the overarching system and do not affect the design beyond the local module (Baldwin and Clark 1999). This independence allows a firm to standardise components and to create product variety. Modularity attracts managerial focus as firms strive to rationalise their product lines and to provide ever-increasing variety at lower cost (Ulrich 1995).

Modularity provides a potential solution to growing complexity and a strategy for organising complex products and processes efficiently. Modular solutions attract managerial focus through the need to align to organisational design (Ulrich 1995) given that different modules need different teams and organisational units to deliver them, e.g. in a modular computer, different peripherals, such as screen, graphics card, and memory storage, are designed from different materials and by different engineering teams. Modularity in management relates to: (i) product design; (ii) production systems; and (iii) organisational design. Product design modularity has implications on the economics of product modularisation and the relationship between product modularity and performance (Campagnolo and Camuffo 2009). There is no pre-determined starting point, that is, the modularisation of the product design does not necessarily come before organisational modularisation (Sanchez and Mahoney 1996).

Many industries today are 'de-integrated', consisting of different firms that each develops one component of a larger product, a module. Hence, modularity changes industry structure from vertically integrated to horizontally integrated structures. Industries evolve due to modularity, which allows ever-increasing specialisation, e.g. in the birth of the automobile industry, car manufacturers were manufacturing the whole car from chassis to windshield, but now they mainly focus on chassis and overall aerodynamics and outsource windshields, seating, even engines. Many industries of complex products evolved into modular architecture, e.g. computer, telecom, and automotive. Design decisions affect the technical characteristics of products and the organisation of production. Modular designs help people divide up the work in tasks. Modularity pushes innovation by allowing for incremental spot innovations on modules only without changing the overall system (Baldwin and Clark 1999).

5.2.2 Platform Thinking

Platforms relate to the proliferation of sharing principles among modular components (Baldwin and Woodard 2009; Cusumano and Gawer 2002). From management scholars, the term platform has been used to denote a design/concept/idea of products, technological systems, and transactions. Wheelwright and Clark (1992) wrote about 'platform products' as a new generation of products for easy modification and substitution. This was followed by 'platform investment', 'platform technologies', and 'platform thinking'. Gawer and Cusumano (2002) wrote about technology strategy 'platform leadership' and how platform groups controlled other firms. In industrial economics, platforms were described as mediators between transactions building upon competitive dynamics (Rochet and Tirole 2003).

The concept of platform relates to the existence of common or reused components and modularity principles, collections of assets, such as products, components, people, knowledge, and the proliferation of sharing principles among the components or assets (Baldwin and Woodard 2009). A platform architecture entails a list of functions, components that perform functions, interfaces, and arrangement of components and descriptions of system operation (Casper and Whitley 2004). Platforms may be found inside firms (internal platforms), across supply chains (supply chain platforms) or as building blocks of industries, also referred to as industry platforms (Gawer and Cusumano 2002). To this end, modular platforms carry implications for organisations and business models on how to organise activities, resources, and partnerships around modular components and business units in the digital economy.

High-tech industries are more likely to have platforms. Microsoft is a technology platform, but there are other types of platforms, such as transaction platforms, for communication, payments, business, and commerce, for example, Amazon and Alibaba. Platforms create new markets by linking a variety of options in modular services or products with customers. Multi-sided platforms, such as Uber or AirBnB, create multiple new markets, e.g. Uber Eats creates three distinct markets of customers, restaurants, and drivers. Platform evolution and orchestration are new concepts and sources of much debate in academic discourse. As platforms are new forms of organisation, there is not a lot of empirical evidence of how they evolve (Cusumano and Gawer 2002). Nevertheless, it is generally accepted that platforms evolve in the opposite manner of products: platform value appreciates through use, whereas product value depreciates through use (Jacobides et al. 2018). At the same time, industries also evolve due to ubiquitous software, as software tools are pervasive in firms and people. Software expertise due to the ubiquity of cheap personal computers has also reached unprecedented levels and availability (Gawer and Cusumano 2002). Advancements in software solutions, e.g. web technologies and cloud solutions, as well as hardware, for example, in accessible devices, such as mobile phones, headsets, and Internet of Things devices, have been paramount in the proliferation of digital technologies. Figure 5.2 summarises the evolution of platforms in technology industries.

5.2.3 Business Models

Business models (BMs) are an integral part of economic behaviour since pre-classical times. Whereas BMs are elusive concepts, they have traditionally focused on business activities and economic growth. Digital technologies act as catalysts for business model experimentation and innovation, opening up new opportunities for organising business activities. Simultaneously, Massa and Tucci (2013) argue that BMs are an important vehicle for innovation and may also be a source of innovation in and of itself.

There are two main roles of business model innovation with regard to digitalisation. First, BMs allow innovative companies to commercialise digital innovation, such as by creating new platforms and new markets. Second, BMs are also a source of innovation in and of itself and a source of competitive advantage, as new organisational structures emerge to address external change other than economic profits, such as responding proactively to digital transformation or solving social and sustainability problems. To this end, firms can compete through new BMs for digital transformation, and a compelling new

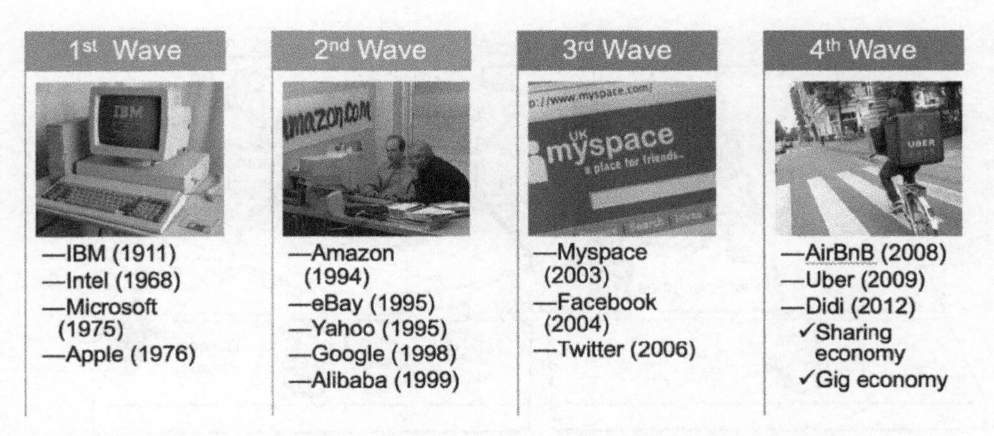

Figure 5.2 Evolution of platforms in the technology industry. *Source:* Rik Myslewski/Wikimedia Commons; The Logo Creative and Franklin Heijnen / Flickr / CC BY-SA 2.0.

business model to address digital transformation may disrupt traditional industries. After all, according to Chesbrough (2007), 'a better BM will beat a better idea or technology'.

5.2.4 Complexity

Complexity is a key challenge in project-based sectors and organising, caused by the high number of interdependencies and relationships among components of a project system. Although a project system with a few components is easy to manage and organise through traditional project management approaches and tools, more complex systems require a network of collaborators coordinated by a large organisation reliant on formal, elaborate, and bureaucratic processes of reporting and control (Brady and Davies 2014).

Adopting system thinking in project management, Tee et al. (2019) illustrated that system complexity relates to numerous interdependences and hierarchical relationships among system component parts. Furthermore, the more complex the system is, the higher the likelihood of information uncertainty, risk, and feedback loops will be, making the task of coordination and the management of projects more difficult (Hobday 1998).

Complexity in management can be defined as relational complexity, that is correlation among system elements, and cognitive complexity, that is limited cognition of system elements (Boisot and Child 1999). Cognitive complexity relates to digital technologies that alter how information is communicated and managed in organisations. Apart from modularity, among various remedies for complexities that have been proposed, digital technologies have been seen as promising solutions. In project-based industries, the digital economy is seen as another remedy for dealing with complexity. In order to organise for digital transformation, the integration of digital technology into all areas of the business is needed and fundamentally changing the ways to organise resources, for example, through open-source digital solutions, how to relate to not only partners, but also platform complementors, and looking at competitors as a potential source of knowledge and value exchange rather than simply antagonists. Figure 5.3 shows the main components of a business model that are transformed by platform thinking in construction.

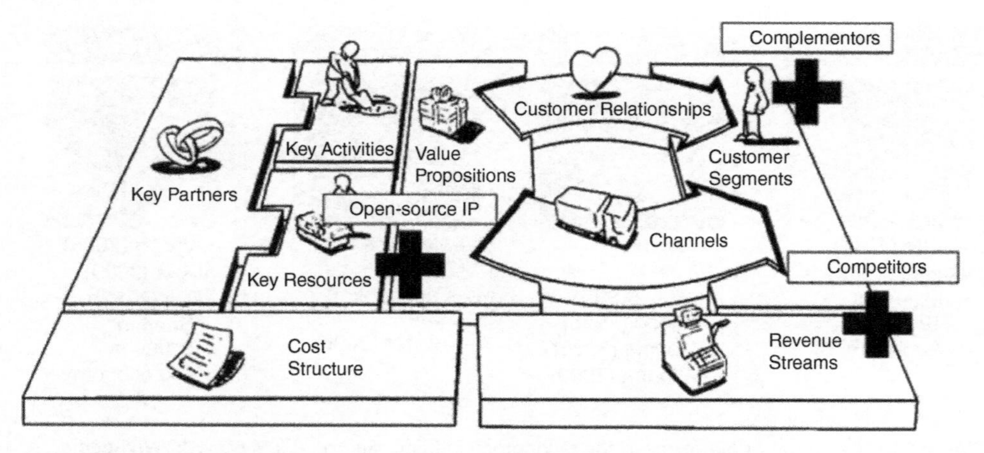

Figure 5.3 The transformation of business models with platform thinking.

The built environment can be seen as a complex ecosystem: it is produced largely by project-based organisations in conditions of complexity (Brady and Davies 2014). The interdependencies between firms are, therefore, high (with their strength being dependent on numerous factors). While digital platforms offer a means to deal with this complexity, the digital innovations underpinning them are combinatorial in nature (Yoo et al. 2010). Digital innovations, by definition, combine physical and digital components to produce novel products (Yoo et al. 2010). Studies of digital innovations in construction show they are unbounded, moving across traditional organisational boundaries (Harty 2005). Boland et al.'s study demonstrates this movement well, showing the 'wakes of innovation' that flow across supply chains of firms working on built environment projects in response to digital innovations initiated in the lead firm (architect Frank Gehry's firm) (Boland et al. 2007). The need to look beyond traditional firm boundaries in considering organising for digitalisation is, therefore, particularly pronounced.

5.2.5 Business Ecosystems

Business ecosystems are underpinned by platform thinking. Such ecosystems differ to other business constellations, such as markets, hierarches, alliances, and supply chains, in aspects including governance, leadership, and coordination mechanisms (Jacobides et al. 2018). An early definition of business ecosystems identifies them as: *'An economic community supported by a foundation of interacting organizations and individuals.'* (Moore 1996, p. 279). Recent years have seen studies of the business ecosystem model in construction, which expand on this view, focusing on their ability to correlate innovation with new value creation (Pulkka et al. 2016) and their central role in enabling Information & Communication Technologies (ICT) adoption and effective national policymaking (Aksenova et al. 2019).

Critically, business ecosystems are value-creating mechanisms. Business ecosystems focus on firms, stakeholders, and the environment: an ecosystem produces goods and services of value to customers who are themselves members of the ecosystem. The nature of the relationships between ecosystem members influences the value created, as described in the service-dominant logic perspective (Vargo 2009). This relationship includes the need for a co-evolution of capabilities and roles across its members.

Extant research shows that leadership is a key variable factor in firm's journeys towards digital transformation (e.g. Morgan and Papadonikolaki 2022). It is also critical in establishing a successful business ecosystem. Platform leadership is defined as: 'companies that drive industrywide innovation for an evolving system of separately developed pieces of technology' (Cusumano and Gawer 2002, p. 52). If platform leadership is an essential component of successful business ecosystems, its role is wider, but incorporates traditional leadership challenges. In addition to traditional firm leadership roles, platform leaders have three other main 'levers' to draw on (Cusumano and Gawer 2002). First, they need to define the scope of the company, addressing the classic transaction cost economics dilemma as to whether to make or buy (Williamson 1979). Second, they need to design the platform (i.e. what parties are going to be involved and how much modularity they want). Their focus is on value creation; thus, the emphasis is on outputs rather than contracts. Third, they need to manage relationships with external complementors: as platforms do not have the hierarchical model to provide coordination, platform leaders play a key governance role in establishing the 'rules' or codes of conduct for members of the ecosystem (Gawer and Cusumano 2014). These complementors (assets, services, and products) that are key to creating successful ecosystems are a key strategic mechanism that platform leaders should orchestrate. In successful ecosystems, leaders strategically invest in complementors to address complex trade-offs (Rietveld et al. 2019).

Of particular interest in considering digital transformation is the model of platform ecosystems. A type of business ecosystem, a platform is a 'hub of value exchanges, coordinating buyers, and sellers through complementary assets, services, and technologies' (Mosca et al. 2020, p. 7). The connection with digital innovations is also strong – for example, platform ecosystems can create digitally enabled workflows, such as contract administration. The co-creation of value in a platform ecosystem is an underpinning principle as members of the ecosystem have a 'strong vested interested in each other's fates' (Rietveld et al. 2019; Vargo and Lusch 2004, 2010).

Figure 5.4 shows clearly the evolution of business ecosystems in construction, and how this evolution has driven the development of value-creating digital innovations (Papadonikolaki and Morgan 2021). This evolution is driven by a rich interplay between some highly influential large projects, through key individuals (actors) and social groups, through rules and institutions, and through digital technologies and socio-technical systems. It can be seen as occurring through four different transition pathways (Geels 2004), namely substitution (–1998), transformation (1998–2011), reconfiguration (2011–2016), and realignment (from 2016 to present). It is interesting to note that the evolutionary process of the firm described in the following section of this chapter occurs during the reconfiguration phase. In this view, change is not only endogenous in firms but is also

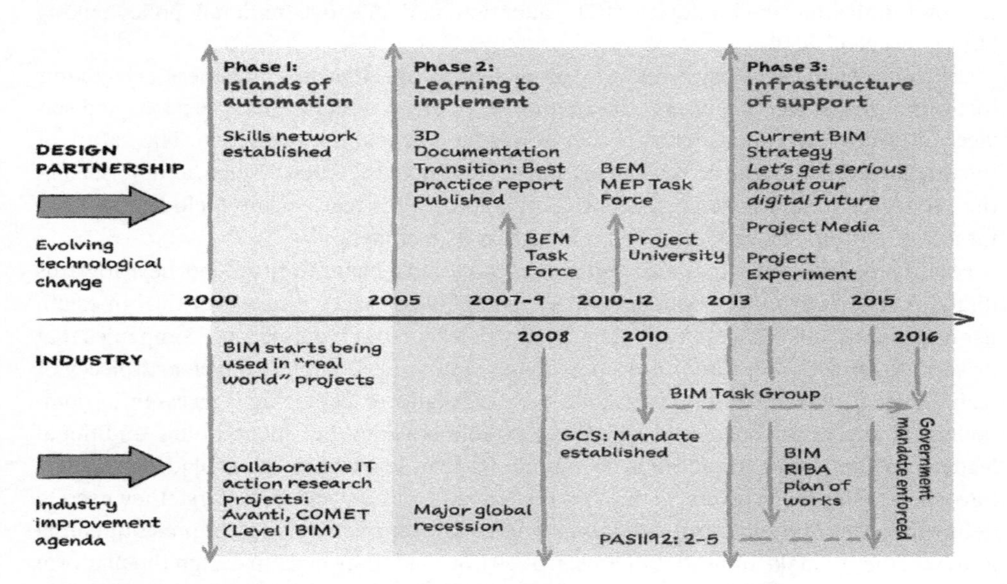

Figure 5.4 Three phases of organising for digitalization. *Source:* Morgan (2019) / Taylor & Francis / CC BY-4.0.

exogenous. Firms act as filters between internal and external environments in driving change, whereby they influence and are influenced by internal and external factors.

5.3 Organising for Digitalisation: New Entrants and Incumbent Firms

The approach to organising for digitalisation differs between new entrants and incumbent firms. As examples in this chapter have described, new entrants are often able to set up with innovative BMs, often enabled by digital technologies, from which they build platform leadership positions. However, for a large set of incumbent firms operating across business sectors, in order to create business ecosystems and organise for digitalisation, they need to change. For many incumbent firms, the magnitude and far-reaching effects of this change are often underestimated. A character in Lewis Carroll's famous novel *Through the Looking Glass* describes this well:

> *'Now, here, you see, it takes all the running you can do, to keep in the same place. If you want to get somewhere else, you must run at least twice as fast as that!'*
>
> *(Lewis Carroll 1872)*

This insight, or the 'Red Queen' effect, underpins many branches of evolutionary studies. Its relevance to organisations, particularly those operating in the face of substantial digital disruption, is apparent. In order for incumbent firms to survive they need to run hard; to get anywhere different (and become disruptors), they need to run even faster (Voelpel et al. 2005).

This is particularly the case for many firms operating in the built environment, who deal not only with competition coming from traditional sources in the market, but also from new entrants. For such firms, the changes demanded by digital transformation are often challenging (Morgan 2019). Research into one such incumbent professional services firms' efforts to organise for digitalisation was undertaken by building a longitudinal case study exploring different change initiatives undertaken in the firm (Morgan 2019). The firm studied operates at the forefront of knowledge development. Its competitive advantage is founded on its reputation for creativity and innovation. Therefore, the need for it to adopt new technologies effectively was critical, and it can be seen as an exemplar organisation in the construction industry.

Three phases were identified in this longitudinal study covering 15 years, from the first phase that was characterised by isolated 'Islands of Automation' to the middle phase where the firm 'Learnt to Implement', through to the final (also the shortest and most effective) phase where an 'Infrastructure of Support' was created, which, in turn, enabled a firm-wide adoption of a radical technology (see Figure 5.3). A number of key variable factors enabled this transition, namely leadership; a richer approach to skills and training meeting the varying needs across the firm; and a willingness to articulate the 'rules of the game' by providing firm specific guidance, routines and procedures. As Figure 5.5 shows, the focal

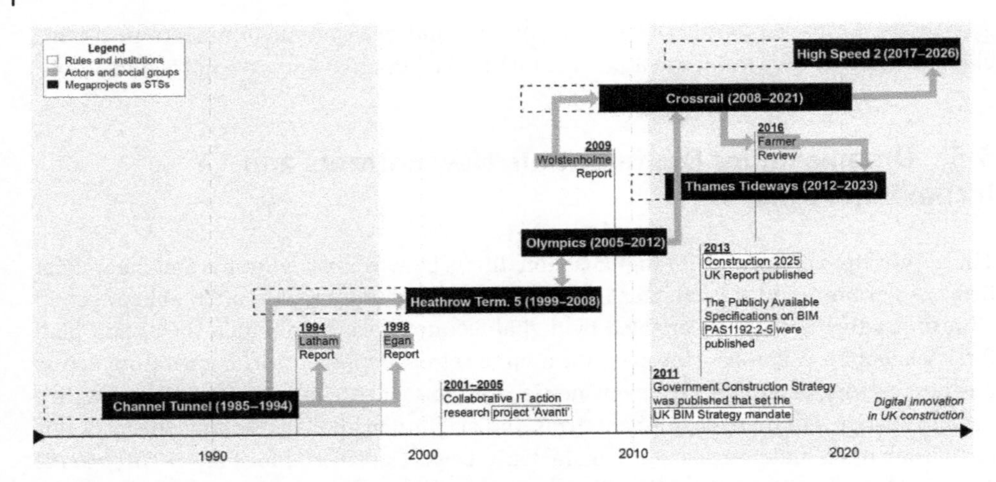

Figure 5.5 The evolution of ecosystems in the UK Built Environment since 1985. *Source:* Adapted from Papadonikolaki and Morgan (2021).

organisation's ecosystem also played a significant role in its organising for digitalisation (this is shown below the timeline). During Phase 3, an escalation in activity across the industry's institutions (including policy bodies and professional associations) is clearly apparent, catalysed through the government mandate and subsequent policy intervention (Morgan 2019).

It is clear from this that digital change requires not only effective organisational leadership, but that external leadership is needed at a wider ecosystem level to drive change. This study suggests that evolutionary approaches to organising for digitalisation necessitate a different view of the firm. They act as a filter between internal and external environments in driving change or the business ecosystem. Following the principles of platform thinking, the consideration of competitors, complementors, and clients is vital as most of the industries are followers in general and certainly not market makers.

5.4 Future States

An important element of organising for digitalisation in the built environment is how we perceive the future. At the same time, platforms are relatively new concepts and form of business organisation, and there is a lot of uncertainty on how they evolve (Cusumano and Gawer 2002). Here, we draw together discussions from the above sections to explore how we in the built environment may use future state thinking to organise for digitalisation, drawing on models of ecosystems, platforms, and modularity to enable this.

As with all of society, the built environment and construction is faced with multiple future states. Dodgson et al. (2013) discuss the importance of innovation process of being 'future ready', for example, in avoiding disruption. These processes prepare organisations for the future by building awareness and responsiveness, for example, changing BMs to deal with disruption in technologies, markets, regulations, and so forth. Chan and

Cooper (2010) posited that 'futures thinking' is emergent and subjective, rather than the objectivised position put forward in many other future focused publications (see, for example, World Economic Forum 2018; McKinsey 2017).

We adopt here Chan and Cooper's stance of a subjective and emergent future, which is shaped by actors. We draw on a methodology that is grounded in this approach and enables individuals to envisage future states that draw on the principles of ecosystems, platforms, and modularity to enable organising for digitalisation.

5.4.1 Future State Founding Principles

Future states bring together modularity, platforms, ecosystems, and other elements into a compelling and cohesive worldview that describes how customers, employees, suppliers, competitors, and others will interact in the future. Future states describe a world that has new features and functions that may not even exist today. It is challenging and exciting and should demand rigorous attention. The core elements of a future state often include the following concepts:

- Future states depict compelling, often counterintuitive, ways of operating. For example, the modern athletic stadium, historically a place of purely public participation, is evolving towards a highly personalised future state. Technologies like location awareness play an important role in being able to assure security, deliver food to seats, redirect foot traffic, and enhance the customer experience. Technologies like Clear Secure, a biometric authentication company, authenticate the individual upon entry, and additional location technologies provide highly accurate location data of customers. The future of event-based attendance is one where individuals have highly personalised experiences in very public places.
- Future states are built on emerging fields. In order to understand if a product or service component is relevant for the future, we must look to the edges of the marketplace – the start-ups and university research facilities where emerging technologies and BMs take shape. We can also look at growth rates within relevant industries for current technologies. These indicators will provide directional signposts to where the relevant components of the future are headed.
- Future states are created in iterations, or launches, that introduce significant new value with each new launch. A future state may have five or more launches to realise the new world. Each new launch should be corrected based upon the feedback of prior launches.
- Future states are designed in a modular fashion in order to clearly understand which components to retain or eliminate based on customer feedback and performance metrics. The modular approach ensures an agile environment and solution pivots as needed for each launch.

5.4.2 The Advantages of Using Future State Thinking in the Built Environment

In traditional design-build environments in the AEC industry, design tends to focus on use: retail, residential, pedestrian and vehicle traffic, and the like. Smart design teams create spaces that encourage sense of place and flow and achieve goals including sustainability, differentiation, and efficiency. While this method has resulted in some notable designs – from the Guggenheim Museum in New York City to the Infinity Tower in Dubai – there are also many design-build efforts that result in a poor use of space, unimaginative designs and leave the inhabitants uninspired.

In contrast to this approach, by building using future state thinking, by starting with the conceived future state, we can expand and contract the elements within the future state in order to improve and simplify the overall design. Multiple future states can compete against each other in the design phase in order to rationalise and justify the best use of the space. Design becomes more than a collection of material and digital assets, instead focusing on the space as a fully formed environment.

Building information modelling (BIM) represents one of the most influential advances in construction innovation in the past decade. Yet, the technologies that form BIM, disruptive as they are by themselves, are not inherently disruptive to the future of the built environment. In fact, any current technology is, by its nature, an incremental solution – a building block to a greater future state. So, we must begin to envision buildings that talk to us, environments that adjust and bend, and intensely personal spaces that compel us by virtue of experience. Design teams must envision these future environments and leverage BIM to support their visions.

In this way, we are always considering the next project as a part of a larger platform. The new building is pushing forward elements and value propositions that advance and adjust our understanding of the future state. We are never designing and building just for that one space, but rather as iteration towards the future. Indeed, the whole definition of modular construction may evolve more to mean that the pieces can be swapped out over time.

Industry disruption has arrived – showing up in the supply chains, demands of customers, new technologies, and contract language. Builders cannot possibly hope to keep up with every new type of concrete or energy saving material. Instead, they need to develop compelling future states, under which all of these newer technologies and infrastructure pieces fit together. Designing with future states in the forefront is a freeing experience because it removes the design team from the addiction of the latest trend. Instead, we know that we are building from a platform towards something of real value to the end user, and in this way, we bring new importance to the work that we do.

5.4.3 Using Future States to Create a Business Ecosystem

Future states often result in a business ecosystem. Some principles that may be useful in the development of future states, ecosystems, and platforms include the following.

1) **Cohesive development teams.** A cohesive development team has purview over the full design and the impact of decisions. Teams must be more than collection of individuals, and should be oriented towards a common goal, have specific roles, and be able to maintain a level of productive disharmony that promotes more vigorous discussion. In this way, trust is formed.
2) **Clear decision criteria.** A typical leadership team makes hundreds of decisions (if not thousands) over the course of a design-build effort. Many decisions are operational in nature; however, a team that lacks clear decision criteria will expend a lot of effort negotiating how to decide. So, setting up decision metrics early in the process is a critical success factor.

3) **Strong pilot practices.** Pilots, prototypes, experiments, and other such activities are an important method for gathering real-world data about what works. Teams who engage in strong pilot practices tend to have faster times to market because they drive faster successes and failures. More 'at bats' creates more data and opportunities for course correction. Where many teams go off track is when the pilot never ends or if a team does not implement the findings.

When it is used well, future state is an approach that can create platforms and ecosystems that help realise the challenges and opportunities implicit in digital transformation, in a range of industry sectors including the built environment. Essentially, it is a human-centred approach and a highly effective method to engage teams work on increasingly complex demands and systemic problems in our built environment. Digital technologies are an increasingly important means of creating such environments.

5.5 Conclusion

In this chapter, we have expanded on modularity, ecosystems, and platforms. The main theoretical contribution of this chapter is to explore these constructs as a means of organising for digitalisation where digital technologies play a central enabling role in creating this means of organising. The role of digital as an enabling mechanism is central: it is the means to an end, rather than the end itself. The relevance of these constructs is particularly applicable in the built environment because of its inherent qualities of complexity and risk. While these constructs underpin a rich stream of research in strategy and business literature more widely, there are many research questions worthy of research, specific to the creation of the built environment.

For example, how are business ecosystems created in our industry? What is the role of the lead firm and how is the governance of the ecosystem managed? How do procurement and contracts support the creation of business ecosystems? We introduce a key pertinent theme for the AEC industry, which is: what are the implications of the ecosystem approach for different ages of firms, focusing on new entrants as opposed to incumbents? Future research may also consider other key variables for firms, such as firm size, and function and position in the supply chain. A final theoretical area of interest is the issue of industry boundaries: because the ecosystem model starts with the customer and the output, this implies that traditional industry boundaries are broken down. What are the implications of this for the existing AEC industry?

Changing industry boundaries are clearly illustrated in the main practical contribution of this chapter. The future state model is output and user focused, particularly in its disruptive state. It presents a possible process, which is enabled by digital technologies for envisioning and realising future challenges and opportunities. It is a user-focused method, which is premised on creating human-centred solutions to our built environment. While future studies have received some attention in construction, they tend to be focused on scenario planning methods (World Economic Focus 2018) or based on expert opinion (Chan and Cooper 2010). In line with the disruptive nature of digital change, we call for more radical approaches to future envisioning in the built environment.

References

Aksenova, G., Kiviniemi, A., Kocaturk, T., and Lejeune, A. (2019). From Finnish AEC knowledge ecosystem to business ecosystem: lessons learned from the national deployment of BIM. *Construction Management and Economics* 37 (6): 317–335.

Baldwin, C.Y. and Clark, K.B. (1999). *Design Rules: The Power of Modularity*. Cambridge, MA: MIT Press.

Baldwin, C.Y. and Woodard, C.J. (2009). The architecture of platforms: a unified view. *Platforms, Markets and Innovation* 32: 19–24.

Boisot, M. and Child, J. (1999). Organizations as adaptive systems in complex environments: the case of China. *Organization Science* 10: 237–252.

Boland, R.J., Lyytinen, K., and Yoo, Y. (2007). Wakes of innovation in project networks: the case of digital 3-D representations in architecture, engineering, and construction. *Organization Science* 18: 631–647.

Brady, T. and Davies, A. (2014). Managing structural and dynamic complexity: a tale of two projects. *Project Management Journal* 45: 21–38.

Campagnolo, D. and Camuffo, A. (2009). What really drives the adoption of modular organizational forms? An institutional perspective from Italian industry-level data. *Industry and Innovation* 16: 291–314.

Carroll, L. (1872). *Through the looking-glass and what Alice found there*. Philadelphia: Henry Altemus Company.

Casper, S. and Whitley, R. (2004). Managing competences in entrepreneurial technology firms: a comparative institutional analysis of Germany, Sweden and the UK. *Research Policy* 33: 89–106.

Chan, P. and Cooper, R. (2010). *Constructing Futures: Industry Leaders and Futures Thinking in Construction*. Wiley.

Chesbrough, H.W. (2007). Business model innovation: it's not just about technology anymore. *Strategy and Leadership* 35: 12–17.

Cusumano, M.A. and Gawer, A. (2002). The elements of platform leadership. *MIT Sloan Management Review* 43 (3): 51.

Dodgson, M., Gann, D. M., & Phillips, N. (Eds.). (2013). *The Oxford handbook of innovation management*. OUP Oxford.

Gann, D. (2000). *Building Innovation: Complex Constructs in a Changing World*. Thomas Telford.

Gawer, A. and Cusumano, M.A. (2002). *Platform Leadership: How Intel, Microsoft, and Cisco Drive Industry Innovation*. Boston, MA: Harvard Business School Press.

Gawer, A. and Cusumano, M.A. (2014). Industry platforms and ecosystem innovation. *Journal of Product Innovation Management* 31 (3): 417–433.

Geels, F.W. (2004). From sectoral systems of innovation to socio-technical systems: Insights about dynamics and change from sociology and institutional theory. *Research policy* 33 (6-7): 897–920.

Glass, J., Bygballe, L., and Hall, D. (2020). Editorial. *Construction Management and Economics*.

Harty, C. (2005). Innovation in construction: a sociology of technology approach. *Building Research & Information* 33 (6): 512–522.

Hobday, M. (1998). Product complexity, innovation and industrial organisation. *Research policy* 26 (6): 689–710.

Jacobides, M.G., Cennamo, C., and Gawer, A. (2018). Towards a theory of ecosystems. *Strategic Management Journal* 39 (8): 2255–2276.

Massa, L. and Tucci, C.L. (2013). Business model innovation. *The Oxford Handbook of Innovation Management* 20 (18): 420–441.

McKinsey (2017). *Reinventing Construction: A Route to Higher Productivity*. McKinsey Global Institute.

Moore, J.F. (1996). *The Death of Competition: Leadership and Strategy in the Age of Business Ecosystems*. New York: Harper Business.

Morgan, B. (2019). Organizing for digitalization through mutual constitution: the case of a design firm. *Construction Management and Economics* 37 (7): 400–417.

Morgan, B. and Papadonikolaki, E. (2022). Digital leadership for the built environment. In: *Industry 4.0 for the Built Environment* (ed. M. Bolpagni, R. Gavina and D. Ribeiro). Springer International Publishing.

Mosca, L., Jones, K., Davies, A. et al. (2020). Platform Thinking for Construction, Transforming Construction Network Plus, Digest Series, No. 2.

Orlikowski, W.J. (2007). Sociomaterial practices: exploring technology at work. *Organization Studies* 28 (9): 1435–1448.

Papadonikolaki, E. and Morgan, B. (2021). Infrastructure megaprojects as enablers of digital innovation transitions. In: *Routledge Handbook of Planning and Management of Global Strategic Infrastructure Projects* (ed. E.G. Ochieng, T. Zoufa and S. Badi). Routledge.

Papadonikolaki, E. and Wamelink, H. (2017). Inter- and intra-organizational conditions for supply chain integration with BIM. *Building Research and Information* 45 (6): 649–664.

Pulkka, L., Ristimäki, M., Rajakallio, K., and Junnila, S. (2016). Applicability and benefits of the ecosystem concept in the construction industry. *Construction Management and Economics* 34 (2): 129–144.

Rietveld, J., Schilling, M.A., and Bellavitis, C. (2019). Platform strategy: managing ecosystem value through selective promotion of complements. *Organization Science* 30 (6): 1232–1251.

Rochet, J.-C. and Tirole, J. (2003). Platform competition in two-sided markets. *Journal of the European Economic Association* 1 (4): 990–1029.

Sanchez, R. and Mahoney, J.T. (1996). Modularity, flexibility, and knowledge management in product and organization design. *Strategic Management Journal* 17 (2): 63–76.

Tee, R., Davies, A., and Whyte, J. (2019). Modular designs and integrating practices: managing collaboration through coordination and cooperation. *Research Policy* 48 (1): 51–61.

Ulrich, K. (1995). The role of product architecture in the manufacturing firm. *Research Policy* 24 (3): 419–440.

Vargo, S.L. (2009). Toward a transcending conceptualization of relationship: a service-dominant logic perspective. *Journal of Business & Industrial Marketing* 24 (5/6): 373–379.

Vargo, S.L. and Lusch, R.F. (2004). Evolving to a new dominant logic for marketing. *Journal of Marketing* 68 (1): 1–17.

Vargo, S.L. and Lusch, R.F. (2010). From repeat patronage to value co-creation in service ecosystems: a transcending conceptualization of relationship. *Journal of Business Market Management* 4 (4): 169–179.

Voelpel, S., Leibold, M., Tekie, E., and von Krogh, G. (2005). Escaping the red queen effect in competitive strategy: sense-testing business models. *European Management Journal* 23 (1): 37–49.

Weick, K. (1979). *The Social Psychology of Organizing*. McGraw Publishing.

Wheelwright, S.C. and Clark, K.B. (1992). *Revolutionizing Product Development: Quantum Leaps in Speed, Efficiency, and Quality*. Simon and Schuster.

Whyte, J. (2010). Taking time to understand: articulating relationships between technologies and organizations. In: *Technology and Organization: Essays in Honour of Joan Woodward*, 217–236. Emerald Group Publishing Limited.

Williamson, O.E. (1979). Transaction-cost economics: the governance of contractual relations. *The Journal of Law and Economics* 22 (2): 233–261.

World Economic Forum (2018). *Shaping the Future of Construction: Future Scenarios and Implications for the Industry*. World Economic Forum.

Yoo, Y., Henfridsson, O., and Lyytinen, K. (2010). Research commentary—the new organizing logic of digital innovation: an agenda for information systems research. *Information Systems Research* 21 (4): 724–735.

6

A Resilience Perspective on Governance for Construction Project Delivery

Nils O.E. Olsson[1] *and Ole Jonny Klakegg*[2]

[1]*Department of Mechanical and Industrial Engineering, Norwegian University of Science and Technology, Norway*
[2]*Department of Civil and Environmental Engineering, Norwegian University of Science and Technology, Norway*

6.1 Introduction

The purpose of this chapter is to discuss project governance from a resilience perspective. We argue that resilience is a key aspect of project governance, and that project governance is a key tool for project resilience. We will discuss the resilience of projects and implications on project governance seen from different project actor perspectives. To do so, we will revisit a selection of railway cases in Norway.

Governance is used in many contexts and on different levels, including the private and public sectors. According to Robichau (2011), the term *governance* is ambiguous, and an agreed definition seems unlikely. In spite of this, the concept is interesting. In this chapter, our main focus is on governance in the public sector, but we draw on some experiences from the private sector as well. Public governance refers to the formal and informal arrangements that determine how public decisions are made and how public actions are carried out, from the perspective of maintaining a country's constitutional values when facing changing problems and environments.

Project governance also has many definitions. In this chapter, we define project governance as: 'a process-oriented system by which projects are strategically directed, integratively managed, and holistically controlled, in an entrepreneurial and reflected way, appropriate to the singular, time-wise limited, interdisciplinary, and complex context of projects.' (Renz 2007, p. 19). Project governance is of special importance for construction project organising because the project level is 'the proof of the pudding', where governance on all levels meet in actual execution.

Project governance can be further divided in several aspects. One aspect is the governance *of* projects, understood as governance on the asset owners' side, to make sure that the organisation selects the right projects and align projects with organisational strategic goals. Governance *in* projects, on the other side, is where governance means are implemented within the project organisation to secure successful delivery and efficiency

Construction Project Organising, First Edition. Edited by Simon Addyman and Hedley Smyth.

in the use of project resources. In multi-organisational teams, governance may also relate to the ownership of the resources that belong to each single organisation and which are used in the project – a subset of corporate governance. It is descriptive of the fact that every team or individual hired into a project is part of an organisation that has its ownership elsewhere (the consultancy, contractor or other supplier firm). This governance aspect concerns ownership to resources, not ownership to the project goals, even though there is a link through risk allocation in contracts. Together, these aspects of governance make up a complex totality. From a resilience perspective, they all need to be addressed.

Resilience is addressed in many scientific areas with different meanings. In technology, it may mean 'bouncing back', such as in material science where resilience is the ability of a material to absorb energy when it is deformed elastically. In business, it can be the positive ability of a system or company to adapt itself to the consequences of a catastrophic failure. This means that resilience can be related to survival. However, in a governance perspective, it is interesting to ask; survival of whom or what? What unit shall survive? And for what purpose?

In the governance of projects, the survival of individual projects may not be the overall objective. In a corporate governance perspective, it is typically the permanent organisation that is the focus object, or the institution as discussed by Söderlund and Sydow (2019). The concept of strategy is important in governance, and the relationship between the organisation and its projects is often discussed. Artto et al. (2008), for example, highlighted the importance of the project's independence from the owner and the strong stakeholder organisations to succeed. If a project becomes misaligned with overall organisational objectives, the survival of the organisation may require adjustments, or even cancellations of individual projects, or possibly rearrangement of the business ecosystem, which form the project context. In this sense, resilience may also be about keeping the project context stable enough to make delivery possible.

From a theoretical perspective, this chapter aims to add a new perspective on governance by focusing on the concept of resilience in a project governance context. We then provide examples of how this has been done in selected railway cases. Key research questions of the chapter are the following:

- What can resilience mean to project governance in different project phases, and seen from different project stakeholders?
- How can the concept of resilience add knowledge and perspective on project governance?

In the following, we give a brief overview of some previous project governance work, followed by a presentation of selected perspectives on resilience in general and what resilience can mean to project management and governance. Following this, we present four Norwegian railway project cases and what survival can mean to such projects. Our focus will be on high-level governance rather than the governance of day-to-day project operations, and we study projects mainly from a client perspective, even though we include interaction between the client and contractors. Finally, we propose a model for perspectives on resilience in different project phases and institutional levels and use it to discuss our results.

6.2 Theoretical Background

6.2.1 Governance

Governance has been an interest to researchers and practitioners for some time, including studies of power structures in government (Stoker 1998). Later, it has developed into a field of study where the focus is on how influence is happening – in society and organisations (Pierre and Peters 2000). Literature on governance often focuses on macro-perspectives at national and regional levels or corporate governance covering private sector organisations. There has also been significant research on governance in the public sector and even the third (not for profit) sector (Phillips and Smith 2011). Of special importance is the project level, and governance of and in projects has become a major part of project management studies over the last 20 years (Miller and Lessard 2000; Pryke 2005; Crawford et al. 2008; Klakegg et al. 2008; Klakegg and Olsson 2010; Ahola et al. 2014; Müller 2017; Unterhitzenberger and Moeller 2021).

The UK Association for Project Management (APM) defines the governance of project management as 'those areas of corporate governance that are specifically related to project activities. Effective governance of project management ensures that an organisation's project portfolio is aligned to the organisation's objectives, is delivered efficiently and is sustainable' (APM 2011, p. 7). This approach to project governance focuses on the way the corporation ensures that project managers and project organisations succeed in supporting the strategic goals of the organisation. This concept is more open to the influence of leadership and the effort of management than the general project governance literature, which often focuses on the structural means and system perspectives. Müller (2017) highlights that it is relevant to view governance as a system of controls, as processes, and as relationships.

6.2.2 Resilience in Different Research Areas

The term resilience has been used in different ways, and in several scientific areas. In ecosystems, Holling and Gunderson (2002) state that ecosystem resilience is the ability to absorb disturbance without inducing 'system changes'. In psychology and health science (Herrman et al. 2011), as well as human development/behavioural science (Masten et al. 2009), resilience is related to positive adaptation in the context of significant challenges. Similarly, in social-ecological systems, Cinner and Barnes (2019) write about resilience as the capacity to tolerate, manage, and adjust to changing social or environmental conditions while retaining key elements of structure, function, and identity. Related to socio-technical systems, Ruth and Goessling-Reisemann (2019) point out that while resilience is typically perceived as desirable, resilience may also hinder development, when change is desired. In those situations, resilience may become counterproductive.

In decision-making, Grafton et al. (2019) discuss resilience management as planning and adaptation actions that influence social-ecological system characteristics. They further elaborate on social-economic resilience, including issues such as resilience of what objects (system, system component or interaction) are being managed and for whom resilience is being managed.

A general summary from a review of publications on resilience is as follows:

- There are several papers on risk management and crisis and disaster management.
- There are many different areas of application: health and safety, environment, natural hazards, climate change, job security, and career development.
- There are some that focus on information systems, organisational change, human factors, and stakeholder participation.

In the following, we will move from reviewing resilience in different scientific areas to be more specific on project management.

6.2.3 Resilience in Projects

In projects, Rahi (2019) proposes a conceptual framework of project resilience to evaluate the ability of projects to deal with disruptive events and enhance their resilience. Further, Rahi (2019, p. 74) refers to Geambasu (2011, p. 133) as 'the first to introduce the concept of project resilience after an empirical study on major infrastructure projects. The author defines it as "1) the project system's ability to restore capacity and continuously adapt to changes 2) to fulfil its objectives in order to continue to function at its fullest possible extent, in spite of threatening critical events".'

Turner and Kutsch (2015) relate resilience to how projects evolve and realign to be able to face disruptive events. Similar perspectives have been researched in projects using the term early warning signs (Hajikazemi et al. 2015). Further to this, Crawford et al. (2013) look at disaster management and projects, while in project risk management, Sabbag (2020) highlights the human factor in project resilience.

6.2.4 Resilience and Flexibility

We noted two schools of resilience definitions and elaborations from our literature review. One school focused on the returning to the original state of a system (such as an ecosystem or organisation). This seems to be based on the (often implicit) assumption that the original state was in equilibrium, optimal, or at least desirable. The other resilience school opens for the flexibility of the system as a means of survival, and resilience does not necessarily mean to strive for the return to exactly the original situation. From this, we summarise that resilience involves finding a new way forward.

The terms flexibility and resilience share several characteristics. Both are responses to the shock or impulses that reside within uncertainty. However, there are also differences. Many, though not all, definitions and elaborations on resilience tend to highlight the desire to return to the situation before the system was exposed to shock. In a project management context, this is illustrated by for example Kutsch et al. (2015), where incidents drive projects off track, causing deviations and poor project performance. If corrective actions are not taken, projects may not survive, i.e. being terminated. It is a tendency in the project resilience literature to indicate that the objective of such corrective action is to return to the plan. As a potential alternative, flexibility represents a strategy in projects to adapt to changes (Olsson 2006), which is a critical part of project organising and formal re-organisation for execution. Flexibility can also increase the value of a project. This means that flexibility does

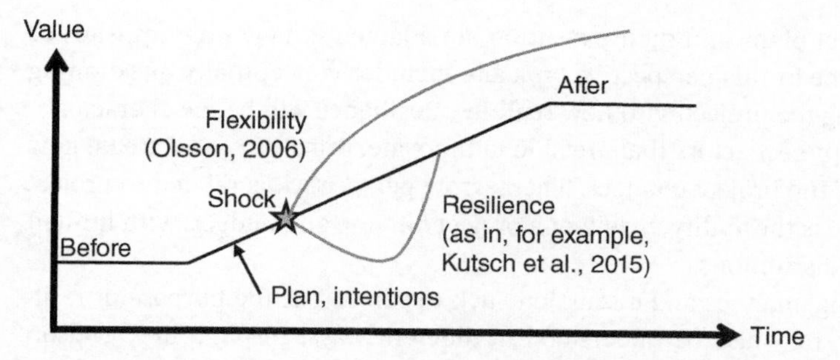

Figure 6.1 Resilience and flexibility in projects as response to shock. Source: Adapted from Kutsch et al. (2015).

not necessarily aim at returning to the original plan, but to work on an adjusted plan, and can aim at an improved project objective. This improved plan represents added value that may manifest as improved quality, added benefits, learning, and new best practice. Figure 6.1 illustrates this. As pointed out by Cooper (2014) and Szentes and Eriksson (2016), flexibility is a key characteristic of successful project organisation from the front end and onwards, but the foundation for such organisational flexibility is laid in the front end.

Figure 6.1 illustrates that the concepts of resilience and flexibility as used in this chapter are both related to how an organisation responds to a shock or change induced stress. The general understanding of resilience focusses on the ability to survive the shock and get back on its feet; flexibility includes the potential to use the current situation (the shock) as leverage to do something positive.

6.3 Reflections on Project Governance, Resilience, and Survival

As shown in previous sections, the resilience concept is used in a wide array of disciplines, including project management, and the understanding of the concept points to resilience as related to survival, recovering from damaging impact and coping with stress. In a project management context, damaging impact can be issues such as new project pre-requisites related to, for example, funding, stakeholder priorities, new policies or strategies, or other issues that cause major changes in context or objectives, or the effects of poor estimates, inadequate project preparations, or critical accidents.

Projects are temporary by definition, and they have a defined life cycle. Temporary project organisations are established and then dissolved when the project delivers its result to mainstream operations in house, or the client or owner where provision is outsourced. Related to concepts of resilience, this is different from most management literature where the studied organisations are typically defined as permanent. In fact, much of the resilience thinking is about avoiding organisations being temporary, to secure organisational survival.

Project resilience can be seen through both a wide and narrow perspective. A wide perspective includes different types of adoption. As for project survival, resilience then looks similar to the practice described by Miller and Lessard (2000), where projects are shaped in episodes or stages. Project plans and their execution are adapted to new pre-requisites or moving goals. Governance in this perspective typically includes conceptually questioning the project and realigning the project with new realities. Resilience will be the characteristic of organisations and project actors that are able to innovate, transform, and create new futures as the context of the project changes. The narrow perspective is related to project delivery, where resilience is the ability to deliver a project on time and budget, with limited change of scope despite disruptions.

We will in the following analyses and discussion stick to survival as the purpose of resilience. Survival, however, may also be understood in different ways. Inspired by Davidson (2010), we suggest the immediate – survive the blast (survival), the secondary – getting back on your feet (adaptation), and third – find another, potentially better, future (transformation).

From a governance perspective, Grafton et al. (2019) ask a number of relevant questions, including survival of whom? What unit shall survive? Related to project governance, some alternative perspectives on this may be:

- the project manager (career)
- the project (the purpose of the intervention)
- the programme (consisting of several projects)
- the organisation (including its portfolio of projects and business as usual)
- the ecosystem/project context/surroundings (the environment or society)

Projects are intentional interventions to change (improve) the client organisation's position in society and business. In case of a shock to the organisation as a system, the project is not given. Projects may be adjusted, delayed, postponed, or even cancelled, depending on the contractual status, if it is for a greater good, such as the survival of the organisation. Our following study will focus on the client organisation, which for railways typically is the infrastructure management organisation. At a programme or portfolio level, projects may be changed or cancelled if the project is revised for a range of reasons. Such adjustments may be triggered by political decisions, policy changes, economic shifts, or other influences, often rooted well outside the projects themselves.

Based on the initial literature study and these reflections, we define the analytical framework, as follows. In line with Cinner and Barnes (2019), we adopt the understanding that resilience is the capacity to tolerate, absorb, cope with, and adjust to change, with a focus on survival.

We will look for answers to the following questions:

- Survival of whom? (key actors).
- Adaptation to what? (key sources of disturbance and stress).
- Transformation of what? (necessary changes and their intended outcome).

In the following paragraphs, we will look into some practical, real-life examples of cases and situations where these questions become relevant.

6.4 Project Cases and Implications on Governance

The issues discussed in this chapter are illustrated with four cases from railway projects in Norway. We look at 'before' decision to finance, 'after' such decision, and the non-survival of the project. We chose railway projects because they illustrate a challenging governance situation for large public projects. Railway investments are large and have long time horizons. They also involve safety-critical systems, and the organisations involved, thus, often need different types of safety-related licences. Railway projects are typically high profile with many stakeholders, including enthusiasts who promote this transport mode. In the cases presented in the following sections, we illustrate examples of resilience and survival and its relationship with governance in the context of the overall railway organisational structure in Norway.

The railway sector in Norway has been through a stepwise new public management inspired split up (Ryggvik 2016). The former public enterprise Norges Statsbaner (NSB) was split into a train operator, an infrastructure manager (now called 'Bane NOR'), and a safety inspectorate in 1996. A further split up of the railway sector was effective from 2017. Competitive bidding for subsidised traffic is ongoing, and new train operators have entered both the freight and passenger markets. In terms of governance, railway and other transport modes sit under the Ministry of Transport, and a new Railway Directorate serves as an interface between the ministry and Bane NOR.

6.4.1 Before Decision – Staying Alive Long Enough to Get Funding

6.4.1.1 Nord-Norge Line

The Northern Norway line is a proposed extension of the Nordland line to the north from Fauske to connect to the Ofot line in Narvik and continuation to Tromsø and a possible side-line to Harstad. The line will pass challenging geography with high mountains and deep fjords, and there are few urban settlements along the line. The idea of linking northern Norway with the rest of the country has been on the table since the late nineteenth century. When looking at a railway map of Scandinavia, the possible line appears as an obvious missing link.

There have been several conceptual studies of Nord-Norge line, including reports in 1982, 1992, 2011, and 2019. Recently, the line has received considerable attention again on the political level. In 2019, the Norwegian Railway Directorate presented an analysis that estimated the costs of a fully developed line at around NOK 130 billion (13 billion Euro) and concludedthatitwillprobablynotbesocio-economicallyprofitable(Jernbanedirektoratet2019). Following this initiative, a wider consideration of the whole transport system in the north of Norway is underway, including the option to build the Nord-Norge line.

Organisational level:

- Survival of whom? The responsible agency Bane NOR.
- Adaptation to what? Pressure from strong promoters to deliver a railway, despite not being a social-economic viable solution.
- Transformation of what? The transport system in northern Norway – and in the perspective of this chapter the railway system in particular.

Project level:

- Survival of whom? Economic actors in northern Norway.
- Adaptation to what? Need for economic stimulus to develop the region.
- Transformation of what? Economic drivers in the region by adding new investments in infrastructure.

The idea has proved remarkably resilient over a century, and particularly over the last four decades, surviving the reorganisations of the railway sector. If history repeats itself, the line may eventually be built if the idea is survives long enough.

6.4.2 After Decision – Staying Relevant and Delivery Through Execution

6.4.2.1 Venjar-Langset

The ongoing railway project Venjar-Langset north of Oslo expands existing double track sections by converting a singletrack line to a double track (Bane NOR 2021). The total cost of the project is approximately 7 billion NOK (0.7 billion Euro). The project includes several complex technical components, of which the Minnevika bridge is of particular interest in this story as it will become Norway's longest railway bridge at 836 m.

To illustrate the resilience in the execution phase, we focus on how the project handled the COVID-19 situation. The contractor with responsibility for building the Minnevika bridge is an international contractor with all its workforce based in other countries in Europe. When COVID-19 hit and travelling between countries became banned, the contractor faced immediate and severe consequences. For the project, this was a real and emergent crisis. Without special provisions, the contractor would have no chance of delivering as planned and according to contract. Bane NOR, as the client, understood they needed to actively help the contractor to achieve exemptions from the general ban on travel across borders.

Bane NOR decided to negotiate with the authorities to gain acceptance for solutions that were in line with COVID-19 regulations for society critical personnel for inbound travel. At potential risk for being held accountable in the media and on a political level for any coronavirus outbreak connected to the project, Bane NOR chose to go the extra mile to help their contractor. All rules for testing against coronavirus, such as a quarantine period upon arrival, had to be followed. Bane NOR as client accepted the time and cost risk for these extra provisions to make sure that the contractor was able to keep up the progress, and thus, the possibility for successful delivery was kept alive.

Organisational level:

- Survival of whom? The contractor, and the reputation of Bane NOR as responsible owner.
- Adaptation to what? COVID-19 pandemic restrictions on travel and transport.
- Transformation of what? Priorities and procedures for transport of materials, equipment, and staff across borders.

Project level:

- Survival of whom? Key roles held by individuals in client and contractor organisations.
- Adaptation to what? Lack of materials, equipment, and staff due to travel and transport restrictions.
- Transformation of what? Time and cost risk to cover extra project procedures and work conditions.

In hindsight, the measures have proven successful despite the client having to accommodate extra costs and some delays. The contractor has kept all regulations and personnel, and materials and equipment have been able to cross borders. The procedures for inbound travel have worked, and the contractor is still on plan for successful delivery.

6.4.2.2 Gardermoen Line

The construction of the Gardermoen line was an integrated part of the decision to build Oslo's new airport Gardermoen, to have an Airport Express train from Oslo to the new airport (Olsson 2006). The line is not only used by the Airport Express trains, but also regional, long-distance, and some local trains. A separate company (NSB Gardermobanen AS) was established in 1992 to plan and build the new line. It was an objective of the Gardermoen line that the Airport Express train should have 50% market share of personnel transport to and from the airport (Government 1991). The airport and new railway line opened in 1998. The Airport Express train had a market share of 33% (JBV Utredning 2004), which is high by international standards. The operation is very reliable with a consistently better performance than other Norwegian trains. The market share for Airport Express trains has recently decreased, because more travellers use other trains on the Gardermoen line to get to the airport. However, The Gardermoen line serves its purpose as a climate-friendly transport alternative and has highly satisfied users.

The decision to integrate the railway line as a part of the airport was key to understand its rapid planning and decision process. The airport was not subject to question; Oslo and Norway needed a new central airport. The railway was more controversial, so by linking the two investments (the airport and the Gardermoen line), the approval of the railway investment was secured.

During construction, the Gardermoen line suffered from severe cost overruns. One reason was connecting the new line to the existing national railway system (a major change to the project that opened up greater benefits). Other reasons included water leaks in a long tunnel and a safety crisis due to the use of poisonous chemicals as a consequence of efforts to seal the leaks. Shortly after its completion, the infrastructure (and related department) was transferred to the governmental infrastructure manager, and the company NSB Gardermobanen was continued as a railway undertaking, renamed to Flytoget and operating the Airport Express trains.

Organisational level:

- Survival of whom? The responsible company (NSB Gardermobanen AS).
- Adaptation to what? Political pressure to contribute to the wider railway network and transport system in society, and to open the same day as the airport.
- Transformation of what? Area of responsibility, business model, and organisational structure of the company.

Project level:

- Survival of whom? The responsible company (NSB Gardermobanen AS).
- Adaptation to what? Serious technical problems causing delay and cost escalation.
- Transformation of what? Plans, working conditions and organisational structure. The Government had to bail out the company by financing most of the extra cost.

As a railway line in operation, the Gardermoen line has shown great results. The Airport Express trains have better punctuality (between 92% and 97%) than other train services and significantly better customer satisfaction, being best in Norway for several years (Flytogets 2019).

6.4.3 Non-Survival of Norwegian Railway Projects

6.4.3.1 High-Speed Railway

High-speed railway has been discussed in Norway for some time. In September 2010, the Railway authority (Jernbaneverket at the time) awarded several contracts for the evaluation of Norwegian long-distance high-speed routes. A final report was submitted in 2012 (Jernbaneverket 2012). The report advised against building long-distance high-speed lines in Norway. A realistic cost estimate (of 1 trillion NOK, or 0.1 trillion Euro) was one of the key reasons for the conclusion. The conclusion paved the way for refocusing Norwegian railway planning, to direct investments for expanding the double track sections further out from Oslo. These expansions, known as InterCity, could be seen as an initial step towards future high-speed investments.

Organisational level:

- Survival of whom? The responsible agency (Bane NOR – previously Jernbaneverket).
- Adaptation to what? Pressure from promoters of railway as a sustainable alternative to air traffic on one hand and the economic realities on the other hand.
- Transformation of what? The priorities among transport modes – potential shift to support railway and reduce air transport.

Project level:

- Survival of whom? Promoting organisations.
- Adaptation to what? Solid documentation that the high-speed railway idea was not viable as a solution for most of the Norwegian transport system and geographical area.
- Transformation of what? The business model of promoting organisations, the scope of transport to be included, and the geographical area.

This case illustrates the resilience concept very well. The original idea of a high-speed railway network in Norway was alive for a long time supported by climate change arguments and ambitions to revolutionise passenger transport in Norway. When the final blow came in the form of solid documentation that this was uneconomic, the idea was developed into promoting potentially more realistic, but still important railway investments.

6.5 Discussion

6.5.1 Different Approaches to Resilience in Different Project Phases

To summarise our observations, we make a distinction between the organisational perspective as well as governance *of* and governance *in* projects and use a simplified project life cycle with three phases – Before decision, Execution and Operation – as shown in Table 6.1.

The governance of projects before a final funding decision is focused on ensuring that it is the right project to be funded, and that the project is based on the right concept. This is well known in the front-end management of projects and project portfolio management. However, in a governance of projects perspective, resilience, such as survival of all project ideas, is not necessarily positive or desirable from the view of the project owner. The corporate governance perspective illustrates the organisational level that focuses on the survival and development of the organisation and the role of the resource owner (either at client or supplier side). Before projects are finally decided upon, governance in projects will be focused on securing funding and support for the particular projects.

During execution, the owner perspective on resilience focuses upon keeping the project in line with strategies and goals for the permanent organisation. Projects that drift off from the path or do not adjust when goals are changed become undesirable and, thus, potentially require the client to import new risks. The project perspective on resilience is keeping focus on delivering the predefined goals. This is referred to as *governance* in projects in Table 6.1. However, this can become adjusted to deliver on moving targets as referred to in the governance of projects in Table 6.1. This expresses a key point about resilience – projects need to be flexible and resilient in the face of a changing environment.

6.5.2 Reflections on the Norwegian Railway Cases

For both the Nord-Norge line and Norwegian high-speed railway lines project governance prior to funding includes questioning the viability of the project, while there is pressure from those who do not need to pay to keep the idea of the project alive. Domestic high-speed lines have not been a hot topic since the high-speed evaluation of 2012, but the work did serve as a trigger for the InterCity project. The death of one project paved the way for another, even though the InterCity project was recently put on hold due to cost escalations.

The Gardermoen line illustrates other aspects of project resilience. A controversial and expensive investment was brought through to realisation. The Gardermoen case demonstrated Gardermobanen's ability to bounce back from severe overrun in execution. The legal entity NSB Gardermobanen AS would have been bankrupt had it not been relieved of its debt, but the construction project survived long enough to be finished, and in operation it is still a success story. Of the studied railway projects, Gardermobanen is the only case where we can study operational experiences. Related to governance in projects, the Airport

Table 6.1 Resilience approaches in different project phases and institutional levels.

Governance perspective	Institutional level	Project life cycle		
		Before decision	**Execution**	**Operation**
Corporate governance	Organisation	*Resource owner:* Create possibilities for continued business.	*Resource owner:* Avoid allocating too much risk and resource overload.	*Resource owner:* Learning and developing new capabilities.
Governance *of* projects	Organisation	*Asset owner:* Questioning profitability, securing value creating decisions.	*Asset owner:* Aligning to organisational goals.	*Asset owner:* Utilising and adjusting the asset for enhanced stakeholder benefits.
Governance *in* projects	Project	*Promoter:* Secure financing.	*Project manager:* Deliver on goals.	*Operators:* Doing business for the long-term survival.

Express train has been commercially successful after the financial restructuring. As for the governance of the Gardermobanen in operation, it has served as a major inspiration for the Norwegian railway sector, being the first high-speed line with good operational performance. The project also provided important learning for following railway projects (Ryggvik 2016).

The recent case of Venjar-Langset illustrates how the parties needed to collaborate to handle the crisis and keep the project on track during hardship that no one had imagined before the project started. The story illustrates how resilience is not the result of one actor's preparations and responses, but the result of joint effort and actions that depended on tough decisions on both sides of the contract (and the national authorities). Crisis looks different from different angles, responses depend on many parties, and the consequences (appropriate allocation of risk) must be shared fairly among the involved parties. Mutual respect, understanding, and information sharing also support the ability to detect, interpret, and respond to changes and shocks. This creates resilience and, thus, survival in execution.

In relation to Figure 6.1, we noted that flexibility and resilience share several characteristics, but there are also differences. In a governance perspective, organising a project is an important process from the front end. However, we find that the surviving railway projects also had the ability to take advantage of contextual changes. The surviving projects not only strived to be resilient in the meaning of returning to plan but were flexible enough to develop new organisational structures. This was the case for Gardermobanen in particular but also to the Nord-Norge line. This points to the need for governance to be flexible when faced with shocks, and this flexibility in governance, in turn, enables the project to be resilient and survive.

The Venjar-Langset project had established a governance structure that served the project well when it was challenged by the COVID-19 situation. This highlights the general importance of the front end, and that organising is an important process in the front end to build a platform for resilience and survival.

6.6 Conclusions

We have looked at the described cases through governance and resilience lenses. In particular, we have used the metaphor of survival as our fundamental interpretation of resilience and its relationship to corporate governance and governance *of* and *in* the projects. Our first research question was what resilience can mean to project governance in general, in different project phases, and seen from different project stakeholders (internal and external).

We found that survival can mean different things for both organisations and projects in different phases, and in different perspectives. It is not obvious that the survival of a project is the overall objective in the governance of projects; good governance can kill projects for the greater good of the organisation. However, we also saw in the studied projects that there was a strong drive to keep projects alive. Resilience was shown to include the possibility of simply being strong and persistent in the face of resistance, or completely the opposite – to be flexible to an extent that includes not only changing its structure, business model, or risk allocation, but in some cases even the project identity.

The concept of project resilience has many organisational aspects, including who is responsible, who defines strategies, and who brings them into life. The basis for resilience includes formally defining what decision-making arenas and entities are legitimate, together with relevant requirements for documentation. In Norway, these are highly centralised political and administrative structures. Accommodating for the public input and influences is regulated in the planning legislation. Process resilience depends on who is allowed to influence decisions, on what levels and at what stages, formally and informally. Transparency is one of the strongest means of building resilience.

The second question asked is related to how the concept of resilience can add knowledge and perspective on project governance. We found that resilience is different as seen with a focus on process, on output or outcome. The disparity is more than the difference in time perspective:

- Focusing on the *process* gives resilience a 'here and now' nature close to early warning signs (ability to detect and interpret signs), risk management (ability to avoid, correct, or live with risk), or even crisis management (the art of intuitively doing the right thing).
- Focusing on *output* gives resilience a nature that is directed at delivery of predefined results and goals that lie in the immediate future under the project's control. This includes project planning and control (the art of systematic balancing resources and goal-directed activity), and the ability to keep the context stable enough to let the project succeed.
- Focusing on *outcome* gives resilience its most fundamental nature – about survival in the long run. Most research on resilience has this focus and looks at ecosystems, open socio-economic technical systems, and other phenomena without time limits, but with identifiable thresholds where the shock may become beyond repair. Projects have this perspective as well, but as temporary interventions to create value and make improvements that last. The object of survival may not be the projects themselves, but the organisations and stakeholders that benefit from projects.

The general organisational structure of the Norwegian railway sector has been in a more or less constant flux in the period of the studied projects, from an integrated railway organisation prior to 1996, to the highly fragmented situation at the time of writing. It may be argued that the fragmentation gives one sort of flexibility and resilience by limiting responsibilities and influence – there are interfaces that can take up some forces and shelter other parts of the system. Each unit specialises and learns to live with its limited, but clear mandate. On the other hand, the use of a rather fragmented structure also points out each unit as responsible and requiring effort to stay and prove itself accountable. So, in this perspective, it seems the system becomes less flexible and less resilient.

This triggers more general questions about what really happens in social–technical–economic systems when the structure gets changed: When there are more interfaces, and the requirements across those interfaces are stricter than the 'all for one and one for all' monolithic structure and culture it came from in the 1900s – does it make the system more – or less – resilient? The signals found in the case stories above are not conclusive. The answer can possibly be both.

Value creation in the long run might represent much stronger incentives for prioritising capability building and strengthening reputation (building resilience), than short-term financial gain from quick fixes and immediate transactions (firefighting). We hope to have illustrated that resilience adds an interesting perspective on project governance and construction project organising more generally. When it comes to survival, project governance may have the objective of project survival, but it may also include the questioning of individual projects.

References

Ahola, T., Ruuska, I., Artto, K., and Kujala, J. (2014). What is project governance and what are its origins? *International Journal of Project Management* 32 (8): 1321–1332.

APM (2011). *Directing Change; a Guide to Governance of Project Management*, 2e. Princes Risborough, UK: Association for Project Management.

Artto, K., Kujala, J., Dietrich, P., and Martinsuo, M. (2008). What is project strategy? *International Journal of Project Management* 26 (1): 4–12.

Bane NOR, (2021). Venjar-Langset project: https: //http://www.banenor.no/eidsvoll last accessed December 30, 2021.

Cinner, J.E. and Barnes, M.L. (2019). Social dimensions of resilience in social-ecological systems. *One Earth* 1 (1): 51–56.

Cooper, R.G. (2014). What's next?: After stage-gate. *Research-Technology Management* 57 (1): 20–31.

Crawford, L., Cooke-Davies, T., Hobbs, B. et al. (2008). Governance and support in the sponsoring of projects and programs. *Project Management Journal* 39 (1): 43–55.

Crawford, L., Langston, C., and Bajracharya, B. (2013). Participatory project management for improved disaster resilience. *International Journal of Disaster Resilience in the Built Environment* 4 (3): 317–333.

Davidson, D.J. (2010). The applicability of the concept of resilience to social systems: some sources of optimism and nagging doubts. *Society & Natural Resources* 23 (12): 1135–1149.

Flytoget (2019). Servicekvalitet i Flytoget 2019. https://flytoget.no/globalassets/rapporter/servicekvalitet/rapport-for-servicekvalitet-2019.pdf Last accessed March 13, 2022

Geambasu, G., (2011). Expect the Unexpected: An Exploratory Study on the Conditions and Factors Driving the Resilience of Infrastructure Projects. PhD Thesis: École Polytechnique Fédérale de Lausanne, Switzerland, Lausanne.

Government (1991). St.prp. nr 90 1991-92 [New Main Airport for Oslo at Gardermoen] Ministry of Transport.

Grafton, R.Q., Doyen, L., Béné, C. et al. (2019). Realizing resilience for decision-making. *Nature Sustainability* 2 (10): 907–913.

Hajikazemi, S., Andersen, B., and Klakegg, O.J. (2015). Barriers against effective responses to early warning signs in projects. *International Journal of Project Management* 33 (5): 1068–1083.

Herrman, H., Stewart, D.E., Diaz-Granados, N. et al. (2011). What is resilience? *The Canadian Journal of Psychiatry* 56 (5): 258–265.

Holling, C.S. and Gunderson, L.H. (2002). *Resilience and Adaptive Cycles*. Washington, DC: Island Press.

JBV Utredning. (2004). Utviklingen i persontrafikken på jernbanen. En analyse av årsakene til nedgangen i trafikken i perioden 2000-2003, Jernbaneverket, Oslo, September 2004.

Jernbanedirektoratet (2019) Ny jernbane Fauske – Tromsø (Nord-Norgebanen) Oppdatert kunnskapsgrunnlag, Dokument nr.21 007 105

Jernbaneverket (2012). Oppsummering: Hovedkonklusjoner. Available at: https://www.banenor.no/Prosjekter/hoyhastighetsutredningen/Nyhetsarkiv/Oppsummering-Hovedkonklusjoner Last accessed July 10, 2021

Klakegg, O.J. and Olsson, N.O.E. (2010). An empirical illustration of public project ownership. *International Journal of Project Organization and management* 2 (1): 16–39.

Klakegg, O.J., Williams, T., Magnussen, O.M., and Glasspool, H. (2008). Governance frameworks for public project development and estimation. *Project Management Journal* 39 (S1): S27–S42.

Kutsch, E., Hall, M., and Turner, N. (2015). *Project Resilience*. New York, NY: Routledge.

Masten, A.S., Cutuli, J.J., Herbers, J.E., and Reed, M.-G.J. (2009). Resilience in development. In: *Oxford Handbook of Positive Psychology* (ed. S.J. Lopez and C.R. Snyder). Oxford: Oxford University Press.

Miller, R. and Lessard, D. (2000). *The Strategic Management of Large Engineering Projects, Shaping Institutions*. Cambridge, USA: The MIT Press.

Müller, R. (2017). Organzational project governance. In: *Governance and Governmentality for Projects: Enablers, Practice and Consequences* (ed. R. Müller), 11–24. New York, NY: Routledge.

Olsson, N.O.E. (2006). Management of flexibility in projects. *International Journal of Project Management* 24 (1): 66–74.

Phillips, S. and Smith, S.R. (ed.) (2011). *Governance and Regulation in the Third Sector: International Perspectives*. Routledge.

Pierre, J. and Peters, P.B.G. (2000). *Governance, Politics and the State*. London, UK: Macmillan.

Pryke, S.D. (2005). Towards a social network theory of project governance. *Construction Management and Economics* 23 (9): 927–939.

Rahi, K. (2019). Project resilience: a conceptual framework. *International Journal of Information Systems and Project Management* 7 (1): 69–83.

Renz, P.S. (2007). *Project Governance: Implementing Corporate Governance and Business Ethics in Non-Profit Organisations*. Heidelberg, Germany: Physica-Verlag.

Robichau, R.W. (2011). The mosaic of governance: creating a picture with definitions, theories, and debates. *The Policy Studies Journal* 39 (S1): 113–131.

Ruth, H. and Goessling-Reisemann, S. (ed.) (2019). *Handbook on Resilience of Socio-Technical Systems*. UK: Edward Elgar Publishing. Cheltenham.

Ryggvik, H. (2016). Sporskiftet. In: *Jernbaneverket 1996-2016*. Oslo: Press.

Sabbag, P.Y. (2020). The human factor in project risk management and resilience. In: *Operations Management for Social Good. POMS 2018. Springer Proceedings in Business and Economics* (ed. A. Leiras, C. González-Calderón, I. de Brito Junior, et al.). Cham: Springer.

Söderlund, J. and Sydow, J. (2019). Projects and institutions: towards understanding their mutual constitution and dynamics. *International Journal of Project Management* 37 (2): 259–268.

Stoker, G. (1998). Governance as theory: five propositions. *International Social Science Journal* 50 (155): 17–28.

Szentes, H. and Eriksson, P.E. (2016). Paradoxical organizational tensions between control and flexibility. When managing large infrastructure projects. *Journal of Construction Engineering Management* 142 (4): 05015017.

Turner, N. and Kutsch, E. (2015). Project resilience: moving beyond traditional risk management. *PM World Journal* 4 (11).

Unterhitzenberger, C. and Moeller, D. (2021). Fair project governance: an organisational justice approach to project governance. *International Journal of Project Management* 39 (6): 683–696.

Part III

The Firm–Project Interface

The chapters in this part address the organising of the firm and the interface between the firm and the project.

First, Smyth addresses firm organising, in particular the main contractor. The key concepts employed are systems of systems, systems integration through the development of capabilities, and loose coupling of the firm. The lack of investment in managing systems and capabilities is a major constraint upon transformation at the project level to improve performance.

Killen et al. proceed by examining project portfolio management. This has scarcely been examined in construction. They focus upon consultants and contractors, drawing their evidence from both Australian and UK evidence. They find some good strategic and governance practices in construction on a project-by-project basis, yet identify some shortcomings around resource allocation from the firm.

Hedborg provides the fourth chapter by examining project ecologies that include a range of organisational actors involved in urban development. In particular, she addresses project organising across the actors by identifying the routines that emerge and shape the programme of development.

Duryan looks at infrastructure development undertaken by a large public sector organisation, examining how communities of practice emerge and can develop to engage individuals in effective learning and knowledge transfer in ways that span hierarchical and functional boundaries.

7

Organising Construction Firms

Hedley Smyth

The Bartlett School of Sustainable Construction, University College of London, UK

7.1 Introduction

Performance improvements in construction are incremental at best. Despite government and client reports repeatedly calling for improvement, transformation remains in the far distance. This chapter argues a major barrier; perhaps, *the* major barrier to transformation is the organisation of the construction firm (Smyth 2018a). Indeed, many of the barriers faced at the project level have their source and flow from the way that construction firms are organised.

The tier 1 or main contractor provides the primary focus. Despite project management bodies of knowledge (PMI 2021) and construction management standardising and routinising, certain activities continue to prove challenging and sometimes unachievable under the current mode of operating. Projects are organised as 'offline activities' that are loosely coupled from the management of the firm (Dubois and Gadde 2002).

Organising at the firm level to provide maximum support for its projects could reasonably be expected to lead to high levels of internal coordination and cross-functional integration at the project front end (Morris 2013), during execution (Turner 2017), and in post-completion (Locatelli 2021). It would also yield opportunities for greater consistency of execution across projects, as well as yielding opportunities for effective learning from projects to the benefit of future projects (Smyth 2018a). This unresolved challenge of systematically organising the construction firm, especially the main contractor, is needed to effectively support projects over their life cycles and between projects. The current system is insufficiently serving contractors, clients, or other stakeholders.

7.2 What the Literature Says

The historic process to project and construction management has been to employ institutional guidelines to provide routinised and standardised approaches to project

Construction Project Organising, First Edition. Edited by Simon Addyman and Hedley Smyth.

management in teaching (e.g. Wysocki 2021; Newton 2016) and research (e.g. Turner 2017; Shenhar and Dvir 2007; Morris 1994; Morris 2013). The most widely recognised practitioner guidance is PMI's *PMBOK*©, which is in its seventh edition (PMI 2021). There are construction-specific forms of guidance (e.g. RIBA 2020). These types of guidance focus on the project, largely ignoring the management context in which project and construction management take place, specifically the construction firms that set many of the parameters and scope for improving project delivery. Yet, it is the construction firm that is 'permanent' or comparatively stable (Winch et al. 2022; Winch 2014), potentially offering resources and support for change and improve performance.

A more dynamic approach has been influenced by management research (e.g. Turner 2017; Söderlund et al. 2012; Morris and Pinto 2007). This has included what is termed project organising (e.g. Winch et al. 2022; Banahene et al. 2014). Project organising is practice-based providing narratives as to what happens on the ground. While there is literature on project-based organisations (PBOs) (e.g. Winch 2014), much of it refers to large complex projects and megaprojects. In this context, either the client organisation is more directly involved in management, making direct interventions (e.g. Davies et al. 2009; Brady et al. 2007), or a semi-autonomous organisation or independent body is formed as the PBO to manage the specific project (Denicol et al. 2021; Davies and Mackenzie 2014). While innovation and learning take place on the exemplary megaprojects (Davies et al. 2014; Brady and Davies 2004), performance improvement is transferred across the megaproject labour market of leaders and senior managers for such projects rather than through embedding the innovations and lessons in the contracting firms (Smyth 2018a).

7.2.1 Government and Industry Reports and Their Limitations for Performance Improvement

There has been performance improvement in construction, much of which has arisen through product development and prefabrication (Best and Meikle 2015). This leaves site-based activities with the residual and thorny problem areas. Achieving performance improvement remains a challenge on the project and especially onsite.

This challenge has given rise to serial prescriptions to break through and overcome the challenges. Some challenges have been identified as arising from some of the industry structures and organisation (e.g. Bowley 1966; McKinsey 2017). However, government, government-sponsored, and client reports (e.g. Adamson and Pollington 2006; Murray and Langdon 2003; Ive 1995 in the UK, and Walker 1995; PRD 1992 in Australia) have made little difference over successive decades (e.g. Smyth 2010).

The arrival of the Internet has accelerated the publication of such reports (e.g. Construction Leadership Council 2020; Farmer 2016; Wolstenholme 2009). Yet, the rapidity and content of these volumes, plus political and industry impetus behind them, has also made little difference (e.g. Cheung et al. 2005; Walker et al. 2015). The reports are frequently followed by high-volume industry and management rhetoric, which has been termed 'fadism' because improvement has been marginal at best (Green and May 2003).

Green (2011) has offered a critical analysis of the government and industry approach to these reports, analysing the evolving approach of organising projects, in particular the growth of subcontracting. Smyth (2018a) has shown that, essentially, the overall approach of management has not changed in the UK since the Second World War, relying on a transactional business model that serves neither firms nor clients and users effectively. The situation is replicated across other countries (Smyth 2022). The consequence is that such reports do not lead to performance improvement to the firm, or industry as a whole.

There is a need to examine the constraints to performance improvement from the perspective of the firm in its operational context alongside the execution of projects.

7.2.2 Systems of Systems, Systems Integration, and Loose Coupling

Construction projects are situated in extended systems of organisational actors that are complex and feature many management and technical interdependencies (Brady and Davies 2014; Morris and Hough 1987). The organisations comprise of firms that provide the multi-organisational temporary project teams (Cherns and Bryant 1984), the clients and sponsor organisations, and trade and institutional bodies, forming the network of individuals representing their organisations (e.g. Greco et al. 2021; Pryke 2012). The active organisational network members have a series of systems inducing systems of systems in which projects are situated (e.g. Whyte et al. 2019; Brady and Davies 2014). Whyte and Davies state in their chapter that systems of systems are active because they make interventions at different levels to coordinate and integrate actions for and at the project level. The interventions involve technical and management inputs to enable systems integration on projects. The majority of analyses of systems of systems have looked at delivery challenges for large complex projects, megaprojects, and programmes (Davies and Mackenzie 2014; Davies et al. 2009; Prencipe et al. 2003). Many of these projects have demonstrated exemplary construction practices, including innovation (Davies et al. 2014, 2009; Brady and Davies 2004). These interventions emanate from PBOs, typically organised by the main client body that act independently of any construction firm (Davies and Mackenzie 2014; Davies et al. 2004). This has also meant that the organisation is focused upon one project or programme to tailor effective interventions for integrating the project systems both technically and managerially. This type of semi-autonomous PBO is more effective than a main or managing contractor being the primary PBO in bringing together resources and project management. However, main contractors undertaking multiple projects are not necessarily geared to any one project or programme with projects being locationally and socially from the firm; hence, the management is loosely coupled (cf. Dubois and Gadde 2002).

A system-of-systems approach intervenes and coordinates inherently layered organisational and management structures at the firm and project levels. In a PBO role, the main contractor is responsible for managing the firm–project interface, which is critical for project integration. This raises two critical areas for the coordination of management and technical capability so that it is seamlessly, hence systematically, brought together for effective integration at each point and stage along the project life cycle. Indeed, there are three critical elements of this systems integration for effective delivery (Whyte and Davies 2021; Davies and Mackenzie 2014; Davies et al. 2009): (i) boundary spanning within construction firms concerning systems integration across departments and functions; (ii) vertical integration in the management hierarchy; and (iii) supply chain management between the main contractor and its subcontractors

Identifying and mobilising adequate resources to invest in capabilities at the firm level is necessary for systematic integration and effective delivery. This is the role of portfolio management in the firm. However, most of the main contractors minimise the investment of this kind. The consequence is that portfolio management moves down to the project level, that is project portfolio management to complete projects rather than improve performance across projects by investing at the firm's programme management level to achieve a consistent service experience and continuous improvement across projects (Smyth and Wu 2021). The investment shortfall constrains capability development, especially in the management capabilities among clients and contractors, for example for organisational learning and knowledge management (Duryan et al. 2020; Fuentes et al. 2019; Brady and Davies 2004) and delivering value post-completion to the client and users (Locatelli 2021; Fuentes et al. 2019; Zerjav et al. 2018).

The concepts of systems integration and investment in capabilities come into conflict with the management of the firm–project interface because construction projects are loosely coupled from the firm (Dubois and Gadde 2002). From the firm perspective, loose coupling is exacerbated by poor systems integration at the firm–project interface and the lack of investment in supportive capabilities.

Loose coupling takes two forms. First is the locational distance of the construction project because execution is inherently site based. Second, yet linked, is the social distance between management in the main office and management on site. For large projects, site-based management teams may be large and co-located to facilitate integration and reduce loose coupling, yet it can only be as strong at the systems and associated routines in the firm and for each project (see also the chapter by Aaltonen and Turkuläinen).

The underlying reason that the management of systems of systems, systems integration, and the effects of loose coupling are weak is that these concepts conflict with the predominant strategy and business model employed by construction firms.

7.2.3 Transactional and Transformational Business Models

The business model is used to implement the strategy of the firm; the model provides the financial instruments and procedures to help enable strategy realisation during and spanning annual accounting periods, although the precise strategy duration will vary between firms. It also describes the means for identifying sources of revenue and making profits through mobilising resources, managing customers, finance and people, as well as its construction projects.

A strategy may involve change and innovation that is built into the model (e.g. DaSilva and Trkman 2014; Baden-Fuller and Haefliger 2013; Teece 2010). Therefore, certain models are flexible, while others are less accommodating of change. However, it is possible to transition to a more transformational approach (Johnson et al. 2008), if the type of value proposition to be delivered is to substantially developed (Amit and Zott 2012). While certain projects may only require specific responses to deliver the required value proposition, hence confined to the project strategy (Zerjav et al. 2018), there are tactical project responses that have strategic implications for the firm. For example, facilitating learning and knowledge transfer from a major project or programme into the firm for the benefit of parallel and future projects (Duryan et al. 2020; cf. Denicol et al. 2021; Smyth 2018a).

There are strategic issues arising from the market in which firms operate that require changes to the business model. The growing complexity of projects and the increased perceived risk and disruption due to strikes of the labour force led to amendment in the late-1960s to mid-1970s in the UK by increasing subcontracting to the point that main contractors became managers rather than producers. While such changes may solve certain operational issues, they may not transform performance, indeed may exacerbate certain aspects (Green 2011). There are attempts being made to put forward changes (e.g. Jones et al. 2019), but there is little substantive to date (e.g. McKinsey 2017). And yet the current business model is not working well for clients, stakeholders, or contractors in terms of profitability internationally (Smyth 2022). There may be some profit relief at the time of writing for firms as interest rates rise again and surplus working capital is invested short term, but it does not solve the performance problems.

At high level, there are two principal types: transactional and transformational business models. The transactional business model dominates construction firms worldwide. It is driven by bidding regimes and finance management. In manufacturing, the process is to invest in management and technical capabilities to innovate, produce, and then sell products, spreading the costs of investment and overheads across the standardised units produced in the long term. In projects, business development and bidding come first, and costs are largely accounted for on a project-by-project basis, which squeezes the ability to invest in systems and capabilities that support improved performance in order to secure bids, typically on the lowest price or next but lowest bid price. Keeping costs low involves keeping overheads as low as possible, hence eradicating investment in capabilities and effective systems management.

One consequence is that profit margins are very low. The prime means of achieving profitability is using trade credit from suppliers and subcontractors, managing cash flow and in particular delaying payments between receipt of stage payments, and paying the supply chain, hence generating surplus working capital to invest in short-term interest accounts, that is, profit through the return on capital employed (Smyth 2022; Gruneberg and Ive 2000). This contrasts with the mainstream approach of returning a profit on the investments made. The return on investment has scope for being transformational.

Therefore, the result is that construction firms, especially main contractors are lean to the point of corporate anorexia and thus hollowed out (Green et al. 2008; Green and May 2005). Consequently, procedures are developed ad hoc without repeatability, and thus, routines are often partial at best, and project organising is frequently left to individual responsibility, where action is often organised around personal comfort zones (Wells and Smyth 2011) to help manage stressful working environments and tight deadlines that frequently challenge staff well-being (e.g. Sherratt 2016).

The transactional model is acceptable if it works; however, the sector delivers suboptimal value for clients, users, and other stakeholders. The model becomes increasingly challenged as society and other businesses require more complex project solutions to their economic, policy, and social needs. In addition, clients, and especially serial clients, are more sophisticated in the value they demand; clients want excellent and consistent performance problems for which completed projects offer solutions (Smyth 2018b; see also Fuentes et al. 2019).

A transformational business model to improve performance with investment in systems, management, and technical capabilities would improve the quality of content and the delivery or service experience, not simply by improving quality around design, specification, and site operations, but potentially leads to learning through the systems of systems what is needed from the frontend into the back end (e.g. Zerjav et al. 2018) and what improves value outcomes for the completed projects in use (Fuentes et al. 2019; Smyth 2018b).

In the UK, the essence of the transactional business model for the modern construction firm was established in the Second World War in an environment of cost plus and target cost contracts rather than lowest bid competitive bidding (Smyth 2018a). Not only was this more sustainable for main contractors, but profit margins, which were around 15% during post-war reconstruction, were potentially fertile for transition to a more transformational business model, but this opportunity for transition was not subsequently pursued in the sector. Since that time, profit margins have shrunk in the UK to 2% or below currently and into single figures in most other countries (Smyth 2022; Smyth 2018a).

7.2.4 Organising and Reorganising the Construction Firm

Low levels of investment in the construction firm, and specifically main contractors, induce constraints across a range of functions, activities, and for systematic integration. At a general level, the consequences arise between portfolio management and project management (Smyth and Wu 2021; see the chapter by Killen et al.), especially at programme management level to help manage the firm–project interface and overcome some of the issues of loose coupling. Specifically, low investment in management and technical capabilities both for and on projects is a norm (cf. Davies et al. 2016). Management is not about simply having a piece of software loaded onto a platform, but it involves a systematic set of processes to analyse what is relevant where, when, and what is generic and will need integration for all projects and what capabilities are project specific, for example project learning (e.g. Brady and Davies 2004). There are a range of such capabilities, that include developing and embedding systems, and specific activities involving organisational learning, knowledge transfer, client management, and key account management (KAM).

In practice, programme management is present among many sophisticated clients with their own work programmes, yet largely absent or very partial among main contractors and subcontractors; most of the main contractors follow client programmes as a means to secure work rather than invest in their own management capabilities (Smyth and Wu 2021). The main reason for low levels of investment in general and programme management in particular is that it challenges the transactional business model, which strategically dominates management decision-making at board and senior management levels. The increase in investment and overhead costs cannot be assigned to any one project, nor recovered through bidding regimes.

The overall consequence of this is not only the lack of management to manage the systems of systems and systems integration, but it is a general lack of a systematic approach. This is manifested at the firm–project interface, exacerbating the effects of loose coupling. Project teams are left to organise themselves. The consequential lack of organising or non-organising at firm level increases the need for project team organising and self-organising at the system-of-systems level (e.g. Whyte et al. 2019; Pryke et al. 2018). This lack of a systematic approach from the firm leads to different service experiences that clients and their representatives endure between projects managed by the same main contractors (Smyth et al. 2019a).

Non-organising at both the firm and project levels is explored in this book (see the Editorial Chapter by Addyman and Smyth), but suffice to say at this point that better organising at firm level and more consistent project organising would help and would need to be supported by a systematic approach to the integration of management activities through functional boundary spanning, in the firm hierarchy and at the firm–project interface. Non-organising also occurs in project teams within the actors and between the actors because of the transactional business model. Low-cost bidding ensures that there are insufficient resources for the construction and project management to effectively organise their operations, resulting in a considerable amount of activity being left to individuals to take responsibility (e.g. Smyth et al. 2019a), constraining the emergence of shared norms to induce effective routines (Eisenhardt and Martin 2000; Schein 1996; see the chapter by Addyman).

The extent of the poor and partial organising and dysfunctional organising that arises due to the low investment and lack of coherent management can be explored through empirical findings.

7.3 Methodology and Methods

This chapter draws upon a series of publications over recent years. The aim is to indicate practices that clearly show a lack of systems integration and poor capability development in the systems of systems and in particular within main contractors. These practices variously illustrate combinations of weak and partial communication and miscommunication, unaligned activities, action, and inaction. Such shortfalls are not only due to the presence of incomplete systems and partial practices, but also where the systems are related to or embedded in the systems of systems among the key stakeholders. The focus is largely upon the supply-side main contractor, but that lack of systems integration inevitably affects other actors in the systems of systems, especially clients and subcontractors in the main contractor supply chain.

Methodologically, each of the research studies is qualitative and interpretative (e.g. Denzin 2002). The studies all draw upon data from semi-structured interviews among key actor organisations, primarily although not exclusively main contractors. The reason for employing an interpretative methodology is because of the strengths of going beyond what, where, and when questions to address why and how questions. This is particularly important because these findings, in the context of addressing the arguments set out and points made in the literature review, concern the gap between theory and practice, where there is a shortfall in practice showing an area for specific improvement, and where there is a conflict or contradiction between theoretical elements where one concept or set of concepts rules and outbids another concept, in particular the strategy and transactional business models of construction firms versus the concepts necessary for mobilisation in order to facilitate performance improvement, especially transformational improvement.

Further, how- and why-type questions help unpack the extent of shared and aligned practices that exist and for these to be further addressed (Eisenhardt and Martin 2000). While the perceptions of the interviewees in all cases may be subjective about the practices, including recall and post-rationalisation on occasions, individuals and teams use these to inform and justify actions, and sometimes ensure their perpetuation (e.g. Sayer 2000) to the point where the dysfunctions become part of the norm, part of the routines, explained in terms of the way things are done (cf. Deal and Kennedy 1983).

The precise methods employed varied between the research studies, depending upon the context, research aims and preferences, and the research teams involved. As the findings and analysis presented here are based upon prior research outputs, any detailed variance of the methods employed is less of a concern than might otherwise be the case because these findings and analysis concern research outcomes in relation to the literature presented rather than the consistency of research outputs.

7.4 A Range of Findings and Analysis

The findings are analysed to draw attention to systems integration issues within the firm and across organisational boundaries. This directly brings in the system-of-systems dimension on the supply side, in particular around firm and project functions and their interaction with the client. Six practice cases were identified. The series of six practices also address the project life cycle, applying the management of projects approach of Morris (2013). The project level brings in loose coupling during delivery, which links back to systems integration.

The first case addresses the practices of the business development function at the front end of the construction project and the lack of integration both at the front end and with commencing execution (Smyth et al. 2019a). Overall, it was found that the contractors examined were all defensive in the interviews. They believed business development functions and subsequent practices yielded a competitive advantage. The defensiveness was echoed in practice, constraining strategy implementation and the tactical co-creation of value through knowledge sharing, innovation, and collaborative working. Most of the reported practices attempted to drive down costs rather than drive performance improvement, which was incremental at best. This epitomises a transactional business model.

One tactical practice was manifest in one of the largest international contractors worldwide. Their business development managers for the UK infrastructure division reported that they did not commit to deliver against certain value elements that clients said will prove beneficial. This was the case even though incorporating the value would give the firm advantage at the bid stage. The reasons given were twofold. The specific value propositions would not be contractually binding, and the business development managers did not have confidence that their project managers would deliver against them, even if it was injected into the bid documentation. This practice shows a lack of integration between the business development, bid management, and project management functions along construction project life cycles. It also demonstrates the power of loose coupling between the firm–project interface (cf. Dubois and Gadde 2002). This point is echoed by Wells and Smyth (2011), whereby project managers organise the projects for which they are responsible according to their comfort zones, yet justify departure from any prescribed project methodology on the basis that they are responsive to client drivers – a clear case of deliberate miscommunication enabled by a lack of systems to address loose coupling. Underlying the problem, business development managers stated that they found it difficult to secure resources from finance management to improve value propositions and prime new sectors for particular projects, even though they had been tasked to do so by senior management (Smyth et al. 2019a).

Procurement management is the focus of the second practice. Procurement has sought to be more collaborative in developing contracts (Mosey 2019; see also the chapter by Addyman). Specifically, they have addressed value (cf. Morris 2013) and use sophisticated models, for example to qualify subcontractors and suppliers, supplemented by key performance indicators to evaluate performance of suppliers and subcontractors that are fed back into the selection process on future projects. There are two issues identified in prior research with this approach. First, almost all the contractors use static key performance indicators, so that suppliers and subcontractors only have to reach threshold criteria to qualify and be selected. This leads to static performance improvement once thresholds are reached (Duryan and Smyth 2019). The second issue concerns boundary spanning to align value required by clients and users with the inputs. The constraint is that the criteria the procurement department selects are developed according to the overall strategy of the firm in a general sense and specifically what the procurement department thinks are important criteria. In one major international contractor, the head of procurement was unaware of the marketing or business development strategy, hence the generic value criteria for particular sectors or market segments, nor liaised with business development and bid management to develop valuable win strategies from the client and user perspective (Smyth et al. 2019a). This is an example of misaligned value propositions due to the absence of a systematic approach to boundary spanning, exacerbated by static internal selection processes.

A third area concerns health and safety (H&S) at the project execution stage. It was found that there is a lack of integration across branch operations of international contractors for H&S. Therefore, construction project managers operate according to the local or national culture and regulatory norms rather than apply a consistent approach (Smyth et al. 2019b). This means that local compliance standards dominate at the expense of a more caring approach to H&S on site and in the office environments. A stark system-of-systems level example is where an international contractor is required to enter into a joint venture with a local contractor in a developing country for bidding and operational purposes. The local contractor could learn from H&S practices in developed countries from the international contractor, but the current lack of systems integration in firms constrains that. A further H&S issue is the conflict between concepts in practice. Concepts to improve H&S are supported by senior management in many large contracting organisations as a top priority alongside profitability. However, implementing financial management concepts dominates the transactional business model squeezing out or trumping H&S onsite. This is in evidence in the decision-making of commercial directors (Roberts et al. 2012) and more broadly where care for the workforce becomes translated as compliance as it cascades down the hierarchy in bidding for projects and driving projects on site where the workforce work long hours under stressful conditions with budget and time constraints (Smyth et al. 2019b; cf. Sherratt 2016). Challenging such matters also challenges the transactional business model, and therefore, not doing so becomes part of the taken-for-granted thinking in construction. This is one reason why statistics on H&S have plateaued in recent years in construction (Smyth et al. 2019b).

Organisational learning and knowledge management specifically is a fourth case. The Egan Report (1998) and the continuous improvement agenda were to provide a major thrust for industry improvement. A review of 150 Demonstration Projects that hailed best practices showed that only three projects were focused upon knowledge management (Smyth 2010). Capturing lessons learned and knowledge transfer onto parallel and future projects is a critical component of driving performance improvement in the industry (Smyth and Longbottom 2005); the absence of a real focus and investment in knowledge transfer capabilities to enable systems integration is a constraint. This absence has system-of-systems implications. Knowledge management is important for some clients in qualifying contractors for projects. Research showed contractors to be active in knowledge transfer by two means. First, IT platforms with software are provided ostensibly for uploading lessons learned; yet, engagement with the platforms was found to be low. This was because project bids neither built in time and cost for doing so nor was contingency funding available at programme management level to facilitate it. Second, post-completion debriefing and submitting reports on lessons learned were too late for the project and parallel projects. Further, relevant knowledge was lost during delivery, and key personnel were changed over the life cycle with inadequate handover to ensure that the knowledge of lessons was handed over; the baton of information and knowledge is dropped, hence failing to reduce the loose coupling effect. Main contractors neither require their supply chains to share nor transfer knowledge to improve operations, and nor did they respond to the client organisation's demand to improve knowledge transfer. The reason provided was that the client constantly changed their minds and investment would be sunk without benefit. They were defensive and adopted an approach described as learned helplessness (Smyth and Duryan 2020).

The case above leads into a fifth case of programme management. Programme management provides a bridge between portfolio management, where resources are allocated in the firm, and the project (Smyth and Wu 2021). Programme management capabilities are one important area of investment to support project activities. They potentially benefit the firm and projects through supporting systems integration and directly improving performance. Programme management is more highly developed among clients with serial projects, for example commercial property developers and public infrastructure clients. One area that has seen some improvement among contractors is client management or KAM. This recognises that clients, as well as projects, need management. A few contractors have developed systems for client management (Smyth and Fitch 2009). A major contractor developed systems to monitor and manage key clients, knowing the potential value of their clients to their own business over the next 10 years. For example, business development endeavoured to inject the type of value required to help develop win strategies around particular bids. However, another major contractor identified KAMs from their senior project managers. The initiative failed as it was not resourced at programme management level, and instead of being client focused, the KAMs were constantly fighting finance departments for funds. The firm had to reinvent the initiative some years later, although implementation remained partial (Smyth et al. 2019a). Whereas knowledge transfer is largely a systems integration issue that reduces loose coupling, this example has system-of-systems implications as it directly affects the service experience received during project delivery and value received across projects and the client base. Construction firms purely allocating resources directly to projects is termed project portfolio management (see the chapter by Killen et al.), hence omitting complementary programme management systems and capabilities (Smyth and Wu 2021).

Finance management and commercial criteria drive the transactional business model, which provides the sixth case concerning the realisation of value post-completion by clients and users. Contractors focus on the means, in particular the inputs and outputs around time–cost–quality/scope (e.g. Atkinson 1999), overlooking project outcomes that matter to clients and users. Examining projects post-completion provides a source of data for feeding back into design and construction to improve the performance of future projects (Fuentes et al. 2019; see also Smyth 2015). This also helps mitigate front-end decision-making that also focuses on the means of execution rather than outcomes (Smyth et al. 2018), hence supporting systems integration at both the system-of-systems and operational levels. Clients, even experienced clients, with serial projects do not necessarily have all the capabilities to be able to feed outcomes back into design and execution, for example public sector clients have particularly lost capabilities since the financial crisis (Winch and Leiringer 2016).

Clients have not always sufficiently organised to secure valuable project outcomes, many of which have lost capabilities too over the last decade or more (Winch and Leiringer 2016). This is part of the service contractors could and arguably should provide as most other sectors seek product and service improvement.

7.5 Conclusions

The findings and analysis involved six areas of practices that are illustrative of dysfunction among main contractors and other organisational actors in the systems of systems (cf. Whyte et al. 2019: see the chapter by Whyte and Davies). The dysfunctions flow from the transactional business model employed by construction firms. The business model of firms constrains performance improvement at the project level. These practices are far from being exceptional. The business model is induced by the failure of the management of construction firms to change over the last 75 years or more (Smyth 2018a; Smyth 2022; see also Green 2011). It is also driven by the variety of client bidding regimes (Hughes et al. 2006). It would be entirely reasonable to maintain the status quo if it was serving both contractors and clients well. However, contractor profit margins have shrunk worldwide over recent decades (Smyth 2022), and clients are receiving suboptimal outcomes from the projects commissioned (Smyth 2018b).

This chapter has argued through the literature review that construction firms, and in particular main contractors, are not organising to facilitate effective systems integration, system-of-systems management, and in the reduction of the loose coupling of projects. Part of the solution is to transition towards a transformational business model, which would have to go hand-in-hand with restructuring procurement in the market (Smyth 2022). A transformational business model can facilitate the transformation of project delivery.

At this point, an exciting agenda for future research and practice could be expected. The excitement is in the differentiation of contractors. Investment to improve performance has to be undertaken incrementally to successfully transition to a transformational approach.

Further investment cannot be on all fronts, but selective to build on strengths, plug gaps, and serve particular type of client and sector, which gives rise to the differentiation and yields competitive advantage. Apart from yielding improved margins in the long term, it will grow the firm because the improvements will help grow market share (Smyth 2022). Research cannot pursue a one-size-fits-all approach, but through normative and prescriptive methods, the options available within the firms and across the networks of interdependent actors can be examined, especially pursued through action and engaged research.

Overall, improving the organisation of the construction firm is a key element to supporting effective project organising and needs to be one key element of transforming construction through improving performance.

References

Adamson, D.M. and Pollington, A. (2006). *Change in the Construction Industry: An Account of the UK Construction Industry Reform Movement 1993–2003*. Abingdon: Routledge.

Amit, R. and Zott, C. (2012). Creating value through business model innovation. *MIT Sloan Management Review* 53 (3): 41–49.

Atkinson, R. (1999). Project management: cost, time and quality, two best guesses and a phenomenon, its time to accept other success criteria. *International Journal of Project Management* 17 (6): 337–342.

Baden-Fuller, C. and Haefliger, S. (2013). Business models and technological innovation. *Long Range Planning* 46 (6): 419–426.

Banahene, K.O., Anvuur, A., and Dainty, A., (2014). Conceptualising Organisational Resilience: An Investigation into Project Organising, Conference Paper Presented at the *30th Annual ARCOM Conference* (1–3 September). Portsmouth.

Best, R. and Meikle, J. (2015). *Measuring Construction: Prices, Output and Productivity*. Abingdon: Routledge.

Bowley, M. (1966). *The British Building Industry: Four Studies in Response and Resistance to Change*. Cambridge: Cambridge University Press.

Brady, T. and Davies, A. (2004). Building project capabilities: from exploratory to exploitative learning. *Organizational Studies* 25 (9): 1601–1621.

Brady, T. and Davies, A. (2014). Managing structural and dynamic complexity: a tale of two projects. *Project Management Journal* 45 (4): 21–38.

Brady, T., Davies, A., Gann, D., and Ruch, H. (2007). Learning to manage mega projects: the case of BAA and Heathrow terminal 5. *Project Perspectives* 24: 32–39.

Cherns, A.B. and Bryant, D.T. (1984). Studying the client's role in construction management. *Construction Management and Economics* 2 (2): 177–184.

Cheung, F.Y.K., Rowlinson, S., Jefferies, M., and Lau, E. (2005). Relationship contracting in Australia. *Journal of Construction Procurement* 11 (2): 123–135.

Construction Leadership Council, (2020). Roadmap to Recovery: an industry recovery plan for the UK construction sector. www.constructionleadershipcouncil.co.uk/wp-content/uploads/2020/06/CLC-Roadmap-to-Recovery-01.06.20.pdf, and updated October, www.constructionleadershipcouncil.co.uk/wp-content/uploads/2020/10/R2R-Status-Report-v5.0-FINAL.pdf (Accessed 17 February 2022).

DaSilva, C.M. and Trkman, P. (2014). Business model: what it is and what it is not. *Long Range Planning* 47 (6): 379–389.

Davies, A. and Mackenzie, I. (2014). Project complexity and systems integration: constructing the London 2012 Olympics and Paralympics games. *International Journal of Project Management* 32 (5): 773–790.

Davies, A., Gann, D., and Douglas, T. (2009). Innovation in megaprojects: systems integration at Heathrow terminal 5. *California Management Review* 51 (2): 101–125.

Davies, A., MacAuley, S., and DeBarro, T. (2014). Making innovation happen in a megaproject: London's Crossrail suburban railway system. *Project Management Journal* 45 (6): 25–37.

Davies, A., Dodgson, M., and Gann, D. (2016). Dynamic capabilities in complex projects: the case of London Heathrow terminal 5. *Project Management Journal* 47 (2): 26–46.

Deal, T.E. and Kennedy, A.A. (1983). Culture: a new look through old lenses. *The Journal of Applied Behavioral Science* 19 (4): 498–505.

Denicol, J., Davies, A., and Pryke, S. (2021). The organisational architecture of megaprojects. *International Journal of Project Management* 39 (4): 339–350.

Denzin, N.K. (2002). The interpretive process. In: *The Qualitative Researcher's Companion* (ed. A.M. Huberman and M.B. Miles), 349–366. Thousand Oaks: Sage.

Dubois, A. and Gadde, L.-E. (2002). The construction industry as a loosely coupled system: implications for productivity and innovation. *Construction Management and Economics* 20 (7): 621–631.

Duryan, M. and Smyth, H.J. (2019). Service design and knowledge management in the construction supply chain for an infrastructure programme. *Built Environment Project and Asset Management* 9 (10): 118–137.

Duryan, M., Smyth, H.J., Roberts, A. et al. (2020). Knowledge transfer for occupational health and safety: cultivating health and safety learning culture in construction firms. *Accident Analysis and Prevention* 459: 105496.

Egan, J., (1998). Rethinking Construction: The Report of Construction Task Force. Report, London: Department of Environment, Transport and the Regions.

Eisenhardt, K.M. and Martin, A.J. (2000). Dynamic capabilities: what are they? *Strategic Management Journal* 21 (10–11): 1105–1121.

Farmer, M. (2016). *Modernise or die: time to decide the industry's future*. Construction leadership council. London. www.constructionleadershipcouncil.co.uk/wp-content/uploads/2016/10/Farmer-Review.pdf (accessed 17 February 2022).

Fuentes, M., Smyth, H.J., and Davies, A. (2019). Co-creation of value outcomes in projects as service provision: a client perspective. *International Journal of Project Management* 37 (5): 696–715.

Greco, M., Grimaldi, M., Locatelli, G., and Serafini, M. (2021). How does open innovation enhance productivity? An exploration in the construction ecosystem. *Technological Forecasting and Social Change* 168: 120740.

Green, S.D. (2011). *Making Sense of Construction Improvement*. Oxford: Wiley-Blackwell.

Green, S.D. and May, S. (2003). Re-engineering construction: going against the grain. *Building Research & Information* 31: 97–106.

Green, S.D. and May, S. (2005). Lean construction: arenas of enactment, models of diffusion and the meaning of 'leanness'. *Building Research & Information* 33 (6): 498–511.

Green, S.D., Harty, C., Elmualim, A.A. et al. (2008). On the discourse of construction competitiveness. *Building Research & Information* 36 (5): 426–435.

Gruneberg, S.L. and Ive, G.J. (2000). *The Economics of the Modern Construction Firm*. Basingstoke: Macmillan.

Hughes, W., Hillebrandt, P.M., Greenwood, D., and Kwawu, W. (2006). *Procurement in the Construction Industry: The Impact and Cost of Alternative Market and Supply Processes*. Oxon: Taylor and Francis.

Ive, G.J. (1995). The client and the construction process: the Latham report in context. In: *Responding to Latham* (ed. S.L. Gruneberg). Ascot: CIOB.

Johnson, M.W., Christensen, C.M., and Kagermann, H. (2008). Reinventing your business model. *Harvard Business Review* 86 (12): 50–59.

Jones, K., Davies, A., Mosca, L. et al. (2019). *Changing business models: implications for construction*. Transforming Construction Network Plus, Digest Series, No. 1.

Locatelli, G. (2021). Is your organization ready for managing "back-end" projects? *IEEE Engineering Management Review* 49 (3): 175–181.

McKinsey (2017). *Reinventing construction: a route to higher productivity*. McKinsey Global Institute, https://www.mckinsey.com/business-functions/operations/our-insights/reinventing-construction-through-a-productivity-revolution#.

Morris, P.W.G. (1994). *The Management of Projects*. London: Thomas Telford.

Morris, P.W.G. (2013). *The Reconstruction of Project Management*. Chichester: Wiley-Blackwell.

Morris, P.W.G. and Hough, G.H. (1987). *The Anatomy of Major Projects: A Study of the Reality of Project Management*. Chichester: Wiley.

Morris, P.W.G. and Pinto, J. (ed.) (2007). *The Wiley Guide to Project, Program and Portfolio Management*. Hoboken: Wiley.

Mosey, D. (2019). *Collaborative Construction Procurement and Improved Value*. Hoboken: Wiley-Blackwell.

Murray, M. and Langdon (ed.) (2003). *Construction Reports 1944–98*. Oxford: Wiley-Blackwell.

Newton, R. (2016). *Project Management Step by Step: How to Plan and Manage a Highly Successful Project*, 2e. Harlow: Pearson.

PMI (2021). *A Guide to the Project Management Body of Knowledge (PMBOK® Guide)*, 7e. Newton: Project Management Institute https://www.pmi.org/pmbok-guide-standards/foundational/pmbok (accessed 19 February 2022).

PRD (1992). Productivity and Performance of General Building Projects in New South Wales. A Report Prepared by Policy and Research Divisions (PRD), Royal Commission into Productivity in the Building Industry in New South Wales, Sydney.

Prencipe, A., Davies, D., and Hobday, M. (2003). *The Business of Systems Integration*. Oxford: Oxford University Press.

Pryke, S.D. (2012). *Social Network Analysis in Construction*. Oxford: Wiley-Blackwell.

Pryke, S.D., Badi, S., Almadhoob, H. et al. (2018). Self-organizing networks in complex infrastructure projects. *Project Management Journal* 49 (2): 18–41.

RIBA (2020). *RIBA Plan of Work*. London: Royal Institute of British Architects.

Roberts, A., Kelsey, J., Smyth, H.J., and Wilson, A. (2012). Health and safety maturity in project business cultures. *International Journal of Managing Projects in Business* 5 (4): 776–803.

Sayer, A. (2000). *Realism and Social Science*. London: Sage.

Schein, E.H. (1996). Culture: the missing concept in organization studies. *Administrative Science Quarterly* 41 (2): 229–240.

Shenhar, A.J. and Dvir, D. (2007). *Reinventing Project Management: The Diamond Approach to Successful Growth and Innovation*. Brighton: Harvard Business Press.

Sherratt, F. (2016). *Unpacking Construction Site Safety*. Oxford: Wiley-Blackwell.

Smyth, H.J. (2010). Construction industry performance improvement programmes: the UK case of demonstration projects in the 'Continuous Improvement' programme. *Construction Management and Economics* 28 (3): 255–270.

Smyth, H.J. (2015). *Relationship Management and the Management of Projects*. Abingdon: Routledge.

Smyth, H.J. (2018a). *Castles in the air? The evolution of British Main contractors*. www.ucl.ac.uk/bartlett/construction/news/2018/feb/castles-air-new-report-charts-management-issues-facing-british-owned-main-contractors (accessed 17 February 2022)

Smyth, H.J. (2018b). Projects as creators of the preconditions for standardized and routinized operations in use. *International Journal of Project Management* 36 (8): 1082–1095.

Smyth, H.J. (2022). Transforming the construction firm? In: *Describing Construction* (ed. J. Meikle and R. Best). Abingdon: Routledge.

Smyth, H.J. and Duryan, M. (2020). Knowledge transfer in supply chains. In: *Successful Construction Supply Chain Management: Concepts and Case Studies* (ed. S.D. Pryke). Chichester: Wiley Blackwell Chapter 14, e-edition.

Smyth, H.J. and Fitch, T. (2009). Application of relationship marketing and management: a large contractor case study. *Construction Management and Economics* 27 (3): 399–410.

Smyth, H.J. and Longbottom, R. (2005). External provision of knowledge management services: the case of the concrete and cement industries. *European Management Journal* 23 (2): 247–259.

Smyth, H.J. and Wu, Y. (2021). *Practice-based study of strategic project portfolio management in the UK construction sector*. www.ucl.ac.uk/bartlett/construction/news/2021/oct/practice-based-study-strategic-project-portfolio-management-uk-construction-sector (accessed 23 February 2022).

Smyth, H.J., Lecoeuvre, L., and Vaesken, P. (2018). Co-creation of value and the project context: towards application on the case of Hinkley Point C Nuclear Power Station. *International Journal of Project Management* 36 (1): 170–183.

Smyth, H.J., Duryan, M., and Kusuma, I. (2019a). Service design for marketing in construction: tactical implementation in business development management. *Built Environment Project and Asset Management* 9 (1): 87–99.

Smyth, H.J., Roberts, A., Duryan, M. et al. (2019b). The contrasting approach of contractors operating in international markets to the management of well-being, occupational health and safety. *Proceedings of the CIB World Building Congress, Constructing Smart Cities* (17–21 June). Hong Kong Polytechnic University, Hong Kong.

Söderlund, J., Morris, P.W.G., and Pinto, J. (ed.) (2012). *The Oxford Handbook of Project Management*. Oxford: Oxford University Press.

Teece, D.J. (2010). Business models, business strategy and innovation. *Long Range Planning* 43 (2–3): 172–194.

Turner, J.R. (ed.) (2017). *Gower Handbook of Project Management*. Aldershot: Gower.

Walker, D.H.T. (1995). An investigation into construction time performance. *Construction Management and Economics* 13 (3): 263–274.

Walker, D.H.T., Harley, J., and Mills, A. (2015). Performance of project alliancing: a digest of infrastructure development from 2008 to 2013. *Construction Economics and Building* 15 (1): 1–18.

Wells, H. and Smyth, H.J. (2011). A service-dominant logic – what service? An evaluation of project management methodologies and project management attitudes in IT/IS project business. Paper presented at *EURAM 2011* (1–4 June). Tallinn.

Whyte, J. and Davies, A. (2021). Reframing systems integration: a process perspective on projects. *Project Management Journal* 52 (3): 237–249.

Whyte, J., Fitzgerald, J., Mayfield, M., Coca, D., Pierce, K. and Shah, N. (2019). Projects as Interventions in Infrastructure Systems-of-Systems. INCOSE International Symposium.

Winch, G.M. (2014). Three domains of project organising. *International Journal of Project Management* 32 (5): 721–731.

Winch, G. and Leiringer, R. (2016). Owner project capabilities for infrastructure development: a review and development of the 'strong owner' concept. *International Journal of Project Management* 34: 271–281.

Winch, G.M., Maytorena-Sanchez, E., and Sergeeva, N. (2022). *Strategic Project Organizing*. Oxford: Oxford University Press.

Wysocki, R.K. (2021). *Effective Project Management: Traditional, Agile, Extreme, Hybrid*, 8e. Indianapolis: Wiley.

Zerjav, V., Edkins, A., and Davies, A. (2018). Project capabilities for operational outcomes in inter-organisational settings: the case of London Heathrow Terminal 2. *International Journal of Project Management* 36: 444–459.

8

Aligning Construction Projects with Strategy

Catherine Killen[1], Shankar Sankaran[2], Stewart Clegg[3] and Hedley Smyth[4]

[1]*School of the Built Environment, University of Technology Sydney, Australia*
[2]*School of the Built Environment, University of Technology Sydney, Australia*
[3]*University of Sydney, John Grill Institute and University of Stavanger Business School, Norway*
[4]*The Bartlett School of Sustainable Construction, University College London, UK and School of the Built Environment, University of Technology Sydney, Australia*

8.1 Introduction

Strategy blends knowledge and capability, strengthening the power to accomplish things. Knowledge is needed for both the concepts with which to imagine a future and the know-how and skills with which to get there. Capabilities, as the *power to* get things done, are needed to implement ideas, visions, and plans. Strategy has a double edge; it is both what is conceivable *and* what is doable – as such it is a *practice*. Hence, strategy is concerned not just with the imagined 'ends' (future desired state or 'visions') but also with how to turn visions into 'reality'. For construction companies running multiple major projects, essential knowledge and capabilities consist of the ability to maintain strategic knowledge of the project processes and their evolution across the range of projects plus a balanced portfolio of projects and their pipeline for the future. The power to accomplish these is critical; each successful project helps build and maintain a construction contracting firms' profile and makes sure its strategic assets are deployed effectively.

Project portfolio management (PPM) is a capability designed to manage multiple projects strategically and holistically. Research repeatedly demonstrates how PPM, tailored to context, improves the effectiveness of senior management decision-making (Martinsuo 2013; Müller et al. 2008) as a strategic capability (Clegg et al. 2018; Kopmann et al. 2017a) in industries developing new products, services, or software (De Reyck et al. 2005; Kester et al. 2011; Killen et al. 2008). Senior decision-makers in the construction industry are under increasing pressure to incorporate longer-term strategic perspectives and holistically consider the environmental and social impacts from construction. Portfolio-level approaches assist with such challenges in other environments; however, there is little research on portfolio-level approaches in construction contracting firms bidding for and delivering

Construction Project Organising, First Edition. Edited by Simon Addyman and Hedley Smyth.

construction projects. We explore the alignment of construction projects with strategy through four case studies and offer insights for the development of PPM capabilities in construction.

8.2 PPM and Strategy

'A portfolio is a collection of programmes, projects, or operations managed as a group to achieve strategic objectives' (PMI 2013, p. 3). PPM adopts principles and the 'portfolio' label from financial portfolio theory (Markowitz 1952); however, formula-based approaches similar to those used in financial contexts have limited application for PPM due to the complexity of project environments (Farrukh et al. 2000; Killen et al. 2008).

Most PPM research occurs in industries associated with innovation and new product development (NPD), including manufactured products, service products, and information technology (IT) development. These traditional areas for PPM research will be referred to as 'NPD environments' in this chapter. Research generally finds that only around half of new products and services succeed, whereas for IT projects, success rates are even lower (Cooper 2019; De Reyck et al. 2005). PPM is associated with improved rates of success through guiding strategic alignment and providing competitive advantage (Kaiser et al. 2015; Killen and Hunt 2010; Martinsuo and Killen 2014). There are no required or standard processes for PPM, although a range of common approaches have been identified (Kester et al. 2011; Lerch and Spieth 2012), including protocols determining project inclusion, and structured formats for presenting data visually for senior executive evaluation and decision-making (Levin 2015; Levine 2005). Tailoring PPM processes for specific contexts delivers the greatest benefit (Crawford et al. 2006; Loch 2000; Martinsuo and Geraldi 2020). PPM decisions strive to allocate limited funding to optimise value from the portfolio while maintaining 'a balance of project types, and [ensuring] that the project portfolio fits with resource capability' (Killen and Hunt 2010, p. 158). Such decisions to approve or adjust funding for projects are made within the organisation, usually as part of a 'stage-gate' process (Cooper 2008).

8.3 PPM and Strategy in Construction

In common with other strategy research, the focus in construction is more on strategy formulation than strategy in practice (Junnonen 1998; Kaiser et al. 2015). The construction industry is fragmented and complex, with high levels of risk and uncertainty. Research in construction environments suggests that portfolio-level management perspectives are used to align projects with strategy, especially through project selection (Kaiser et al. 2015).

For construction contracting firms (referred to as construction firms, hereafter), selecting which projects to pursue (through bidding) is the most important element in determining the project portfolio composition. Profit remains the dominant goal (Bizon-Gorecka and Gorecki 2019), with workforce availability the most cited selection criteria. Alignment between the strategy and selection of projects requires extensive consultation with stakeholders on the development of a strategic vision (Naaranoja et al. 2007).

The role of strategy (Ansari et al. 2015) in aligning with collaboration (Pryke and Smyth 2006; Smyth 2015) is important.

A practice-based study identified organisational structure as driving information flow and portfolio selection criteria (Kaiser et al. 2015). Smyth (2015) emphasises the importance of structural alignment and a focus on strategic information requirements. A staged model for capturing, reporting, and managing information enables past performance to inform decisions regarding the future portfolio (Hurtado et al. 2018). The portfolio of projects must be adjusted as opportunity and risk profiles change (Erzaij et al. 2020) with data-driven 'intelligent PPM' proposed to reduce risk and enhance resource utilisation. To this end, data-driven visualisations employing business analytics software showed positive results in a trial, despite not being adopted (Al-Sulaiti et al. 2021). Multi-project oversight underpins continuous improvement as expectations change, such as a concern for sustainability (Ershadi et al. 2021).

8.4 Method

How construction firms align their portfolios of projects with strategy in practice remains a key question. We explored portfolio-level management approaches through qualitative practice-based research methods. Such methods are suitable for providing insights to guide tailoring PPM for construction (Clegg et al. 2018; Gherardi 2019; Martinsuo and Geraldi 2020). We collected data from four prominent construction firms: two in Australia and two in the United Kingdom (UK). Between four and eight managers from each firm participated in semi-structured interviews regarding their strategy and project processes. We also interviewed five expert practitioners who advise client groups or industry bodies. Table 8.1 provides an overview of the cases and interviews. Three interviews were conducted in person, the rest virtually, due to COVID-19 restrictions. The duration of the interviews ranged from 50 to 90 minutes, with most interviews over an hour long. All four organisations were private firms in the construction industry, ranging from a construction consultancy to tier 1 contractors that are capable of delivering major projects.

Semi-structured interviews designed to explore practices used to align projects with strategy were refined after pilot testing. To ground the research in practice, we focused on what the construction firms do rather than prescribing that interviewees address specific PPM

Table 8.1 Case study data.

Case code	Location	Brief description	No. of interviews
ORGA	Australia	Tier 1 contractor	4
ORGB	Australia	Construction consultancy	4
ORGC	UK	Tier 1 contractor and consultancy	8
ORGD	UK	Tier 1 contractor	8
EX	UK and Australia	Experts – advisers and representatives of industry organisations	5

concepts. All interviews were recorded, transcribed, and analysed to reveal themes that emerged during the interviews.

8.5 Findings Addressing the Question 'How Do Construction Companies Align Projects with Strategy?'

8.5.1 The Nature of Strategic Resource Allocation Decisions in Construction

Most construction firms compete to deliver construction projects by submitting a bid via a tendering process that involves a significant investment that can reach millions of dollars or pounds. As success rates are only 25–30%, such bidding involves a major and risky strategic decision. A bid does not necessarily lead to a project but does commit internal funds. The low success rate requires maintaining a steady flow of bids and adjusting plans when bid decisions are made. The internal strategic decisions about whether to bid for projects shape a construction firm's future portfolio of projects. When bids fail, the investment in bidding is lost; however, some large government clients aim to keep contractors interested and maintain a pool of bids by providing compensation for bid costs. Although many projects are announced many years in advance, their timing is hard to predict, given political influence and contingencies such as elections (Wynn et al. 2021). Maintaining a pipeline of successful bids is, therefore, an inherently uncertain process.

8.5.2 Strategy and Competitive Advantage

Construction firms' attention to strategy has increased as the contemporary competitive environment became more challenging. Some interviewees did not see a large role for strategic positioning, while others felt that strategic processes were essential. All case firms identified profit as the main aim while noting increasingly tight margins. Strategies to target niche areas, such as tunnelling, environmental sustainability capabilities, and modern methods for offsite construction, provide opportunities for differentiation and improved margins. Some case firms employed targeted strategies to enhance market awareness of their strengths and specialty areas. Strategy drives decisions on whether and how to bid on projects. Projects most likely to be pursued are ones that offer opportunities for repeat business while also helping maintain a high reputation, thus enhancing word of mouth recommendations. Enhancing and protecting the firm's reputation and developing relationships with clients and senior industry professionals is a central aspect of each of the case firm's strategies.

Most of the construction firms report at least five years of strategic planning. One firm explicitly adopted a '531' approach: five-year strategy, three-year plans, and one-year budgets. Another explained that the granularity of strategy was higher in the one- to three-year timeframe but that strategic pipeline planning extended for 10–20 years. Our findings suggest that increasingly longer strategic timeframes are now needed to select the right sectors and projects to fit with the firm's capabilities and provide adequate returns. Risk is reduced by careful choice of customers and partners with whom to work. One interviewee noted that the approach to bidding has shifted from a 'whack a mole' approach (hit anything that appears) to a much more selective approach over the past decade. During this time, Australian firms evolved from primarily state-based to national entities, while in both Australia and the UK, overseas-owned firms were increasingly active.

Enhanced strategic positioning increases construction firms' competitiveness. Some firms invest in innovation to develop specialised knowledge and capabilities that enable them to achieve increased margins. Employees' specialist skills enable construction firms to bid for a range of work to balance industry cycles and align capabilities with opportunities. Diverse capabilities enable contractors to conduct a larger percentage of the work and minimise sub-contracting, enhancing profit margins and reducing risks while maintaining the legitimacy of a high reputation, so important for future business success. Firms invest in activities to attract, retain, and upskill employees. Initiatives for shifting industry cultures, practices, and perceptions aim to increase diversity and strengthen the pool of potential employees.

Aligned with the increasing importance of strategic planning and reputation, the construction industry has seen an increase in 'business development' roles responsible for a long-term pipeline of possible opportunities as well as short-term decision processes to determine opportunities to pursue. Despite reporting that strategic considerations and conversations are increasing at higher levels, strategy is not felt to be effectively communicated to employees. The COVID-19 pandemic contributed to this lapse by upending working patterns, interrupting culture and communications.

8.5.3 Multi-Project Management

Perspectives varied between managers that considered the portfolio as 'everything' to those that avoided the term 'portfolio' but still articulated the importance of a higher perspective (akin to a portfolio-level view) in guiding decision-making on short- and long-term opportunities. There was strong evidence of portfolio management approaches in play, notably among the larger firms. The term 'portfolio' tends to be used by those with strategic oversight, responsible for decisions about projects to pursue. Those firms demonstrating portfolio-level vision generally considered projects as a group rather than regarding them individually, aiming to achieve a balance across the portfolio to deliver strategy and manage resources. Portfolio management was also used by some case firms to track key client relationships and their programmes, which was instrumental in developing long-term strategic relationships and enhancing alignment of projects with strategy.

8.5.4 How Is Portfolio Management Achieved in Construction Firms?

The definition of a portfolio is the first step in portfolio management, that is, how projects are grouped and what types of projects are considered in strategic portfolio decisions. The case firms generally grouped projects according to organisation structures or strategic business units (SBUs), based on type of work, geography, or industry segment; for example, one firm had a division for rail projects, whereas others had divisions for geographical areas. Some multi-layered structures were also observed, for example, road and rail within a geographical area. The pipeline of opportunities and projects is generally managed through such structures, often affording a de facto definition of the portfolio, streamlining management, and reducing risk by improving the selection and governance of projects.

Within a given portfolio, construction firms use processes for developing bids using 'stage-gate' (also called 'phase gate') processes in a manner similar to processes used in NPD environments; each stage in the process is followed by a 'gate' where evaluation and decision-making determines whether to proceed to the next stage. Portfolio oversight often informs decision-making at these gates.

Decisions about whether and how to bid on a project follows sequential stages in the case firms. Despite variation, the approach typically includes: (i) scanning the environment for opportunities; (ii) filtering these opportunities to identify the projects offering the best fit with organisational strategy, structure and capabilities; and (iii) developing a business case prior to preparing a bid. If the proposal to bid is approved and funded, three additional stages follow: (iv) development; (v) review; and (vi), submission. A wide range of maturity and formality levels were evident in these processes. Standard templates are generally used to structure the business case and scoring models assist some organisations in rating proposals.

The three larger case firms (ORGA, ORGC, and ORGD) have clearly-defined and tiered systems to determine decision-making authority, with increased accountability and sign-off limits cascading upwards in routinised, standardised, and formalised processes. Decisions for most project stages were based on evaluation and decision-making at the SBU level, with some level of consideration of the local portfolio. Higher-level portfolio perspectives were often incorporated for important resource allocation decisions about whether and how to bid, considering factors such as contribution to reputation, track record in strategic areas, risk management, and resource availability (workforce, specialist skills, and equipment) as well as balancing the geographic spread of projects. The smaller case firm (ORGB) used portfolio-level input less often for decisions on bidding for work. Portfolio-level oversight and balancing was evident, however, in the ways the smaller firm sought to target strategic projects, diversify their work, and match jobs with capability, even when not signified by the term 'portfolio'.

A senior manager is usually responsible for managing the bid process for an SBU and is likely to be instrumental in resource allocation decisions. Formal approval within the organisation ranges from local agreement and sign-off to processes with multiple levels of approval that may involve group decisions at the board level. Although these high-level decision tiers are designed to give the 'go/no go' decision, the case firms reported that by the time proposals reached the higher levels, they were usually successful. The 'go' decision was often contingent on adjustments to some aspects (such as margins or risk levels).

Construction firms are awash with data resulting from digitisation and tools such as building information modelling, but are yet to develop systems capable of translating the data into information that can be used to enhance knowledge and support strategic decision-making. Information systems for operational elements such as safety and environmental sustainability are common, and one firm had systems managing client relationships. Improving use of data for strategic decision-making is a priority. Capabilities range from methods to collate and analyse data via spreadsheets to a relatively advanced bespoke system at one of the firms that enables timely and transparent access to data. Although the data in the bespoke system were relatively current, further development aimed to make data available across the organisation in real time. Methods for visualising information were felt to be in their infancy, with much room to grow and improve. Finally, evidence suggests that although capturing information on 'lessons learned' is recognised by case firms as valuable for improving processes for future projects, it was not done regularly. The difficulties in getting a team together post-project (when many have dispersed and transferred to another project) often prevent effective capture of 'lessons learned', although one firm reported gaining organisational support to bring some of their teams together for this purpose.

8.5.5 Portfolio Decision-Making – Whether and How to Bid

Whether to bid for a project is the most important strategic decision, likely to be made formally and at the portfolio (or SBU) level, incorporating multiple factors. Case firms emphasised the importance of adequate time and resources for bidding. Bidding must be planned thoroughly to determine what resources are required and the bid price margin must be sufficient to make the exercise worthwhile. Case firms were confident they could develop accurate bids and predict margins. Low-margin jobs were not considered; profit was the focus rather than turnover. Interviewees from the case firms, all of which were private companies, felt that public companies were under different pressures, tending to more short-term vision and often seeking to grow and increase turnover rather than focus on profit.

When making decisions on whether to pursue a bid opportunity, the consideration of 'how' to bid is highly strategic: what effort should be placed in the bid process (including relationship building and dialogue) and how aggressively should the work be pursued and priced? Case firms were able to justify lower margins for highly strategic projects, notably those that generated repeat work or provided opportunities to establish a reputation in desired technologies or types of projects. Such decisions were dictated by strategy at the high levels.

8.5.6 Flexibility and Responsiveness

Large projects have long planning horizons, taking many years from the announcement of the project to the release of the tender documents and final decision. Once decisions are made, up-to-date information and rapid, flexible responses are essential. An overwhelming benefit of a portfolio view, including up-to-date information, is the ability to 'see' changes in the situation when a tender decision or a major announcement is made. While a large bid is under review, all bidding firms must be prepared to deliver that project and may hold back on other bids. However, once a tender decision is made, it must be factored into the commitments and ongoing bidding processes in both successful and unsuccessful bidders. For this reason, a good information system that reflects any updates in a timely fashion is essential for informing potential project opportunities, risks, resources, skills, value, and benefits. The case firms identified opportunities for improvement and a desire to enable a proactive response to change without an overly bureaucratic approach that could inhibit their agility.

8.5.7 Repeat Business and Resources

Reputation and repeat business are highly important for the construction firms' success, with one firm estimating 90% of their business was repeat business. A good relationship with the client enhances the outcomes for both parties and working with familiar clients reduces risk. Before bidding for a project, some firms reported scrutinising clients to determine whether there was any conflict of interest as well as the extent to which client's values aligned with theirs on safety and environment. Repeat business was an important factor when choosing where and how to bid. Such decisions not only informed but also shaped strategy; repeat business opportunities influenced the evolution of strategies underpinning the approach to project selection and delivery. Ensuring that projects can be delivered effectively is a major aspect in construction firm decision-making in order to protect their reputation and legitimacy.

An effective delivery of projects depends primarily on the availability of the appropriate resources, at the right place and time. Of most significance is the availability of human resources, specifically the availability of experienced managers and specialists, as well as the availability of specialised material resources such as tunnel boring machinery. Careful consideration is made in portfolio decisions not only to ensure that the committed work does not exceed the availability of resources but also to provide work that deploys and develops resources that might otherwise be idle.

Each case firm cited workforce availability as the main challenge they currently face. An infrastructure construction boom in Australia meant that more work was planned over the next decade than can be met with available human resources. The worldwide COVID-19 pandemic and in the UK, Brexit, have exacerbated the problem. Brexit choked off labour from Europe forcing many workers to return to their home countries after losing residency and visa rights. Government controls instituted during COVID-19 eliminated the option of bringing skilled people into Australia from overseas on specialist visas.

The shortage (and projected shortages in the future) of human resources underlines the importance for the construction industry of diversifying its workforce. The case firms acknowledged that despite concerted efforts and some improvements in the participation of women and other under-represented groups, there remains a need to widen the pool of potential candidates. In addition to strategies to diversify the workforce, the case firms sought to attract, develop, and retain skilled people.

Geography matters in aligning human resources to the project portfolio. The construction firms operate nationally, with work and resources spread across large areas and employee locations and regional offices located thousands of kilometres apart. Gaining work aligned with the locations of staff and acquiring staff in the locations where the work is based is an overarching issue. Firms perform a two-way balancing act that aims for a pipeline of work in the regions where they have the human resources available while seeking to acquire staff in areas where work is active and expected.

8.5.8 Challenges and Performance

Satisfaction with the ability to link projects with strategy was relatively high, although most interviewees cited areas for improvement. Interviewees felt that decision-making processes incorporated strategic considerations well, particularly at early filtering stages. The primary areas cited for improvement included better use of data to inform decision-making, communicating strategy more effectively throughout the organisation and improving the ability to learn and improve via lessons learned.

Universally, the primary challenge facing the construction industry is workforce availability, particularly skilled and experienced employees. Portfolio vision was thought to assist with effective decision-making in balancing resource availability and ensuring resource sufficiency for projects; however, construction firms also noted the need to expand, train, and retain staff. Staff retention was enhanced through positive project experiences and the ability to progress careers, in both of which portfolio approaches play a role. Portfolio visibility can help balance resourcing to ensure that staff are not 'overworked' and to develop skills and careers. Such approaches are currently informal, and there is room to improve.

8.6 Discussion

The four case studies and input from industry experts reveal patterns in how construction firms align projects with strategy. The use of portfolio-level perspectives increasingly provides benefits in using strategy to guide project bidding processes. The findings shed light on decision-making processes in construction and reveal some similarities and differences between PPM in the demand-led construction industry and the NPD environments where different drivers dominate.

Whether or not the case firms referred to 'PPM' or used the term 'portfolio', the findings demonstrate that portfolio-level approaches are instrumental in delivering strategy. Although the success of the bid will determine whether a project is conducted by the construction firm, it is the internal decision about whether to bid for a project that shapes the project portfolio. The nature of the businesses and the ways that projects originate vary considerably between construction and NPD environments; however, the mechanisms for shaping the project portfolio are surprisingly aligned. As with NPD environments, each construction firm's PPM processes are unique and tailored to suit their environment, although common elements are observed. The case firms provided detail of how the decision on whether to bid for a project follows a staged and gated approach that aligned with strategy in a similar manner to NPD environments. Figure 8.1 aligns a typical NPD environment PPM process with a representation of a typical decision-making process used in the case firms. Whereas in NPD environments the major 'go/no go' decision is a decision to develop a product, in construction, the major 'go/no go' decision is whether to bid. In both the cases, the end of the internal decision-making process produces an outcome that is dependent on succeeding in an external context. For NPD environments, the launch of a new product is dependent on customer adoption for success, whereas for construction firms, the success of the bid is determined by the client who issues and awards the tender. In essence, NPD environments produce first and then sell, whereas construction firms sell their services first and then produce.

8.6.1 Strategy and Projects – A Two-Way Relationship

Our findings demonstrate strategy being used as a filter to guide bid decisions at a portfolio level, shaped by the construction project opportunity landscape. The early scanning of opportunities and long planning pipelines influence the evolution of strategies in the case firms.

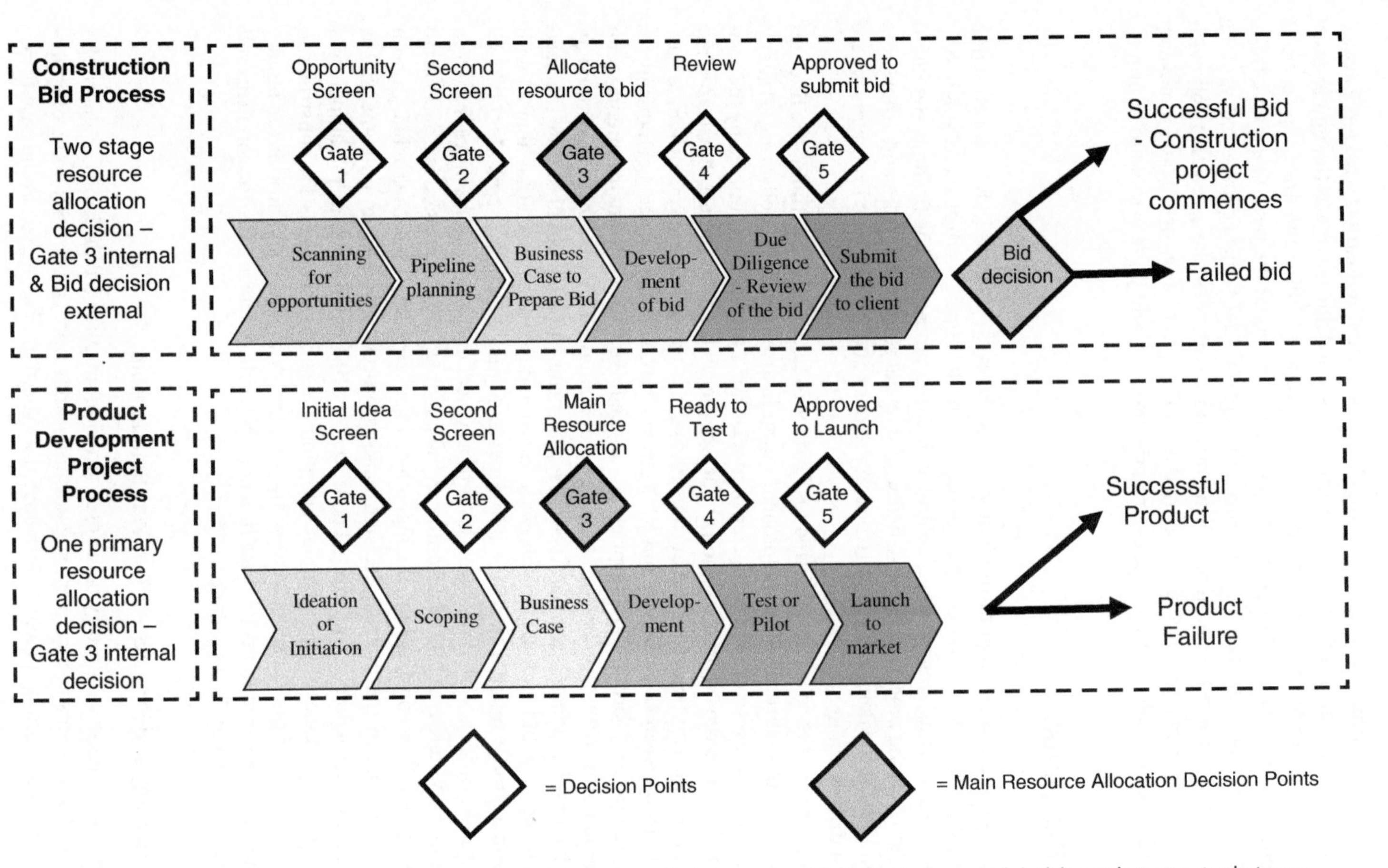

Figure 8.1 Bid development compared with product development – typical processes with defined stages and decision points or gates between each stage.

Interaction between the project portfolio and evolving strategy is also observed in NPD environments. Project portfolio processes inform strategy emergence, while portfolio vision assists in updating strategies to respond to current opportunities (Kopmann et al. 2017b).

8.6.2 Internal and External Decisions Determine the Portfolio

Construction firms reveal project portfolio approaches with many parallels to those used in NPD environments (Behrens and Ernst 2014; McNally et al. 2013). However, construction is a demand-led industry where project opportunities originate externally, whereas NPD projects usually result from internal marketing or technology opportunities. As a result, project portfolio decision processes differ. For construction firms, internal *and* external decisions determine the configuration of the eventual construction portfolio, whereas in NPD environments, portfolio decisions are largely an internal affair. Much of the literature and practitioner guidance on PPM stems from research in NPD environments and focuses on the internal multicriteria decision-making processes used to balance strategic, financial, resource, and other considerations to select projects supporting the best outcomes within a limited budget. However, in construction, there are two distinct decision stages: the decision on whether and how to bid on a project is made internally resulting in a portfolio of bid projects, followed by the client's decision regarding which contracts will be awarded.

The major strategic decisions in construction firms are whether to bid and how to bid. Not every opportunity can be pursued, so organisations must determine which bids provide the best balance of risk and reward. Due to this two-stage decision-making process before a construction project is launched, PPM processes as documented in NPD environments may not immediately seem relevant. In NPD, projects result from processes of ideation, justification, evaluation, and approval, whereas construction projects result from external decisions to award contracts. Less visible are the highly strategic processes for developing the bid portfolio, evaluating opportunities, and deciding when and how to bid.

8.6.3 Flexibility and Responsiveness

The two-stage decision-making process changes the way PPM is done in construction firms, strongly influencing the need for up-to-date information and the ability to adjust and respond quickly. The dynamic complexity and degree of 'unknowns' in the decision-making around where and how to bid are considerable. The advantages of portfolio management in enabling response to environmental changes is well established (Daniel et al. 2014; Killen and Hunt 2010). While NPD environments are often affected by evolving technologies and customer profiles, or by more abrupt changes such as regulatory changes or new competitive offerings, these aspects apply only to some degree in construction. The major change that ripples through the construction industry is when the successful bidder is announced for a major project, prompting a quick reconfiguration by the construction firms.

8.6.4 PPM in Practice in Construction

Despite differing levels of formality and differing terminology, the case firms brought a level of portfolio oversight to the decisions about whether to bid. As with NPD environments

(Kaiser et al. 2015; Killen and Hunt 2010), the structure of the organisation plays a role in portfolio definition and oversight, with structuring and restructuring used to manage resource allocation and risk management in construction firms. The major decision at Gate 3 (see Figure 8.1) is most likely to involve several levels of approval before the 'go/no go' decision to bid is confirmed. In the case firms, strategy was a major filter in the early stages; by Gate 3, strategic alignment was generally clear. The recommendation to request funding to bid for a project was usually done within an SBU and was generally led by the head of that area. In some firms, the decision to bid was essentially made at that level even when higher-level approval (such as via a board) was required; once the process had reached the stage of a formal request for resources to prepare the bid, the request was unlikely to be denied. Therefore, the 'business development' work at the SBU level took on a strong 'portfolio' perspective, including a close watch on the market and the suite of opportunities as well as a good understanding of the resource availability and the need to balance considerations.

The concepts of 'idea generation' and 'the front end' in NPD environments (Martinsuo 2019; Salerno et al. 2015) are replaced by scanning the horizon for opportunities and prospects in construction, not only for projects but also for identifying clients with repeat business. For government projects, many germinate over a long timeframe, especially large projects and megaprojects. Some construction firms maintain a 10-year pipeline vision, with projects noted even beyond that timeframe. As in NPD environments, however, it is common that the three-to-five-year pipeline is the most well defined, and planning is more 'certain' for the shorter term. We found that, generally, the case firms evaluated and updated their plans and strategies in a major review once a year, while also having mechanisms to adjust and refine strategies and plans during the year (usually quarterly or monthly). First pass reviews were sometime done even more frequently, such as in meetings every two weeks at the SBU level.

Some of the tools and methods used in NPD environments, such as business case templates, scoring models, and graphics and visualisations of data, were also evident in the construction firms, however, generally at a lower stage of maturity or formality. Managers in the case firms emphasised the importance of availability, reliability, and transparency of data in alignment with PPM findings from other environments. The visualisation of data is important in fostering sensemaking (Weick 1995). Meaning is constructed by virtue of the senses, primarily through discursive means and visual symbols. Visual symbols perform several functions by affecting cognitive processing, communication, and persuasion. There is growing awareness in management studies regarding the need to consider the interplay of verbal and non-verbal modes of communicating, including the need to consider the visuality of the organisation (Barberá-Tomás et al. 2019; Bell et al. 2014).

Lessons learned is another concept discussed in both NPD and construction environment literature, with both noting benefits and difficulties in conducting project reviews. In both environments, project team members start moving to other projects as the current project winds down, which creates challenges in getting the team together after the project is complete (Duryan and Smyth 2019). These challenges are amplified in construction environments because team members are likely to be geographically dispersed.

8.7 Conclusions and Recommendations

Strategy and its implementation involve the development and application of knowledge and capabilities. PPM is a conceptually important capability. Our exploration of how construction firms align their portfolios of projects with strategy reveals that portfolio-level decision-making approaches are increasingly important and steadily evolving to meet unfolding challenges. We identified three priority areas or sub-capabilities for improving processes for aligning strategy to projects in construction. The suggestions are drawn from findings on PPM in NPD environments while also discussing how PPM in construction can lead improvements to PPM practice to benefit a wide range of industries.

Rethinking: Adopting a portfolio-level perspective will improve the processes for aligning projects with strategy and enhance the ability to respond to change.

Construction firms can benefit from 'rethinking' the processes that underpin success and acknowledging and extending the role of portfolio perspectives. We highlighted the role of portfolio-level perspectives in strategic decision-making concerning whether and how to bid for a project. From the 'front end' processes through to 'launching' the bids, construction firms use processes that align with the stage-gate and PPM processes that provide benefits in NPD environments. A strategic benefit of PPM includes enabling a response to changing conditions, which is an essential capability in construction environments where firms must rapidly reconfigure when major tender decisions are announced. Adopting a PPM mindset and learning from the wide experiences documented in NPD environments may assist construction firms to compete and respond to change.

Data and Strategic Decisions: The ongoing digital transformation of construction environments and the associated wider availability of data present opportunities to make better use of data visualisations to support the decision-making process. This is an area of rapid evolution in both construction and NPD environments, with increasing attention to data capture and presentation. Drawing from experience in NPD environments offers opportunities to construction firms. Tailoring PPM processes to construction could include identifying core visuals that represent the most important factors in the bid decisions and using these visuals to inform decision-making discussions. Data are plentiful, but these are not useful unless systems are available to assist with analysis and communication; investing in systems that assist with presenting data visually is likely to improve decision-making.

The availability of accurate and current data is a challenge in both construction and NPD environments. The types of data are different, and the methods for collecting the data vary widely. In construction, automated methods of tracking progress (Sherafat et al. 2020) represent advanced data capture capabilities that could lead the way for PPM tools that harness data to inform strategic decisions. Sharing experiences across industries will accelerate adoption and translation.

Workforce Planning and Development: Construction firms require advanced consideration of workforce planning and development that may provide insights for PPM in the wider context. Balancing workforce availability and skills is important across industries; however, the challenges specific to construction environments stem from especially high levels of uncertainty and require comprehensive analysis. The low success rate for bids contributes to the uncertainty in workforce planning. Other factors include not only current skills and locations of staff but also how to manage across a series of projects to develop individual and team skills for efficient launch of new projects. Exploring and enhancing portfolio-level approaches for workforce planning and development in construction should provide benefits that can be translated to other environments.

In conclusion, a growing use of PPM concepts is evident in construction firms' processes for aligning projects with strategy. Portfolio-level processes in construction resemble approaches documented in NPD environments but must be tailored to enhance benefits. Wider recognition of the importance of PPM and the ways it is used in the construction industry stand to improve strategic and holistic decision-making capabilities. Such capabilities are especially important as construction firm strategies move beyond a focus on short-term profit to creating a sustainable future, one that is sustainable for both the organisation and the planet.

References

Al-Sulaiti, A., Mansour, M., Al-Yafei, H. et al. (2021). Using Data Analytics and Visualization Dashboard for Engineering, Procurement, and Construction Project's Performance Assessment. 2021 IEEE 8th International Conference on Industrial Engineering and Applications (ICIEA).

Ansari, R., Shakeri, E., and Raddadi, A. (2015). Framework for aligning project management with organizational strategies. *Journal of Management in Engineering* 31 (4): 04014050.

Barberá-Tomás, D., Castelló, I., de Bakker, F.G.A., and Zietsma, C. (2019). Energizing through visuals: how social entrepreneurs use emotion-symbolic work for social change. *Academy of Management Journal* 62: 1789–1817.

Behrens, J. and Ernst, H. (2014). What keeps managers away from a losing course of action? Go/stop decisions in new product development. *Journal of Product Innovation Management* 31 (2): 361–374.

Bell, E., Warren, S., and Schroeder, J. (2014). Introduction: the visual organisation. In: *The Routledge Companion to Visual Organisation* (ed. E. Bell, S. Warren and J. Schroeder), 1–16. Routledge.

Bizon-Gorecka, J. and Gorecki, J. (2019). Establishing the project portfolio management in the construction company. *IOP Conference Series: Materials Science and Engineering* 603: 052014. https://doi.org/10.1088/1757-899x/603/5/052014.

Clegg, S., Killen, C.P., Biesenthal, C., and Sankaran, S. (2018). Practices, projects and portfolios: current research trends and new directions. *International Journal of Project Management* 36 (5): 762–772.

Cooper, R.G. (2008). Perspective: The Stage-Gate® Idea-to-Launch Process – Update, What's New, and Nex Gen Systems*. *Journal of Product Innovation Management* 25 (3): 213–232.

Cooper, R.G. (2019). The drivers of success in new-product development. *Industrial Marketing Management* 76: 36–47.

Crawford, L., Hobbs, B., and Turner, J.R. (2006). Aligning capability with strategy: categorising projects to do the right project and to do them right. *Project Management Journal* 37 (2): 38–50.

Daniel, E.M., Ward, J.M., and Franken, A. (2014). A dynamic capabilities perspective of IS project portfolio management. *The Journal of Strategic Information Systems* 23 (2): 95–111.

De Reyck, B., Grushka-Cockayne, Y., Lockett, M. et al. (2005). The impact of project portfolio management on information technology projects. *International Journal of Project Management* 23 (7): 524–537.

Duryan, M. and Smyth, H. (2019). Service design and knowledge management in the construction supply chain for an infrastructure programme. *Built Environment Project and Asset Management* 9 (10).

Ershadi, M., Jefferies, M., Davis, P., and Mojtahedi, M. (2021). Incorporating environmental sustainability in project portfolio management by construction contractors. *International Journal of Structural and Civil Engineering Research* 10 (3): 123–127.

Erzaij, K.R., Hatem, W.A., and Maula, B.H. (2020). Applying intelligent portfolio management to the evaluation of stalled construction projects. *Open Engineering* 10 (1): 552–562.

Farrukh, C., Phaal, R., Probert, D. et al. (2000). Developing a process for the relative valuation of R&D programmes. *R and D Management* 30 (1): 43–54.

Gherardi, S. (2019). *How to Conduct a Practice-Based Study: Problems and Methods* (2e. Edward Elgar Publishing.

Hurtado, K., Savicky, J., Sullivan, K. et al. (2018). Construction portfolio management model for institutions. Construction Research Congress.

Junnonen (1998). Strategy formation in construction firms. *Engineering Construction and Architectural Management* 5 (2): 107–114.

Kaiser, M.G., El Arbi, F., and Ahlemann, F. (2015). Successful project portfolio management beyond project selection techniques: understanding the role of structural alignment. *International Journal of Project Management* 33 (1): 126–139.

Kester, L., Griffin, A., Hultink, E.J., and Lauche, K. (2011). Exploring portfolio decision-making processes. *Journal of Product Innovation Management* 28 (5): 641–661.

Killen, C.P. and Hunt, R.A. (2010). Dynamic capability through project portfolio management in service and manufacturing industries. *International Journal of Managing Projects in Business* 3 (1): 157–169.

Killen, C.P., Hunt, R.A., and Kleinschmidt, E.J. (2008). Project portfolio management for product innovation. *International Journal of Quality and Reliability Management* 25 (1): 24–38.

Kopmann, J., Kock, A., and Killen, C.P. (2017a). Project portfolio management – the linchpin for strategy formation. In: *Handbook of Organizational Project Management* (ed. S. Sankaran, R. Muller and N. Drouin), 92–105. Cambridge University Press.

Kopmann, J., Kock, A., Killen, C.P., and Gemuenden, H.G. (2017b). The role of project portfolio management in fostering both deliberate and emergent strategy. *International Journal of Project Management* 35 (4): 557–570.

Lerch, M. and Spieth, P. (2012). Innovation project portfolio management: a meta-analysis. *International Journal of Product Development* 16 (1): 77–94.

Levin, G. (ed.) (2015). *Portfolio Management: A Strategic Approach*. CRC Press.

Levine, H.A. (2005). *Project Portfolio Management: A Practical Guide to Selecting Projects, Managing Portfolios, and Maximizing Benefits*. Jossey-Bass, John Wiley distributor.

Loch, C. (2000). Tailoring product development to strategy: case of a European technology manufacturer. *European Management Journal* 18 (3): 246–258.

Markowitz, H. (1952). Portfolio selection. *The Journal of Finance* 7 (1): 77–91.
Martinsuo, M. (2013). Project portfolio management in practice and in context. *International Journal of Project Management* 31 (6): 794–803.
Martinsuo, M. (2019). Strategic value at the front end of a radical innovation program. *Project Management Journal* 50 (4): 431–446.
Martinsuo, M. and Geraldi, J. (2020). Management of project portfolios: relationships of project portfolios with their contexts. *International Journal of Project Management* 38 (7): 441–453.
Martinsuo, M. and Killen, C.P. (2014). Value Management in Project Portfolios: identifying and assessing strategic value. *Project Management Journal* 45 (5): 56–70.
McNally, R.C., Durmuşoğlu, S.S., and Calantone, R.J. (2013). New product portfolio management decisions: antecedents and consequences. *Journal of Product Innovation Management* 30 (2): 245–261.
Müller, R., Martinsuo, M., and Blomquist, T. (2008). Project portfolio control and portfolio management performance in different contexts. *Project Management Journal* 39 (3): 28–42.
Naaranoja, M., Haapalainen, P., and Lonka, H. (2007). Strategic management tools in projects case construction project. *International Journal of Project Management* 25 (7): 659–665.
PMI (2013). *The Standard for Portfolio Management*, 3e. Newtown Square, PA: Project Management Institute, Inc.
Pryke, S. and Smyth, H. (2006). Scoping a relationship approach to the management of complex projects in theory and practice. In: *The management of complex projects: a relationship approach* (ed. S. Pryke and H. Smyth), 21–45. Blackwell, Oxford.
Salerno, M.S., Gomes, L.A.d.V., Silva, D.O.d. et al. (2015). Innovation processes: which process for which project? *Technovation* 35: 59–70.
Sherafat, B., Ahn, C.R., Akhavian, R. et al. (2020). Automated methods for activity recognition of construction workers and equipment: state-of-the-art review. *Journal of Construction Engineering and Management* 146 (6): 1–19.
Smyth, H. (2015). *Relationship Management and the Management of Projects*. Routledge, Abingdon.
Weick, K.E. (1995). *Sensemaking in Organizations*. Sage.
Wynn, C., Smith, L., and Killen, C.P. (2021). How power influences behavior in project management. A theory of planned behavior perspective. *Project Management Journal* 52 (6): 607–621.

9

Urban Development Project Ecologies – An Organisational Routines Perspective

Susanna Hedborg

Department of Real Estate and Construction Management, KTH Royal Institute of Technology, Stockholm, Sweden

9.1 Introduction

Construction project organising is no longer just about planning, designing, and constructing within the single construction project. On the back of the continuous trend of urbanisation and increasing sustainability requirements, construction projects are now performed in more complex project contexts, as they rarely happen in isolation from other projects. From a project management perspective, it is well established that projects do not happen in a vacuum, but rather we need to take history, context, and future expectations into account in order to understand our organised society (Blomquist et al. 2010; Engwall 2003). In other words, we need to understand what is going on around one single project and how this influences the project itself. One example of such inter-project interdependencies is construction projects in urban development districts.

When we look beyond the single project, construction projects performed within an urban development district are seen as being performed within a project ecology (Hedborg et al. 2020), that is, in parallel and sequentially, where participants move between projects and, therefore, have past experiences with other actors and expectations to be part of upcoming projects (Grabher 2004). Using the lens of project ecologies, the focus is on understanding interdependencies being both inter-project and inter-organisational (Söderlund 2004). This is highly relevant for the development and construction industry as projects are almost exclusively inter-organisational and, in many instances, such as in urban development districts, have critical inter-project interdependencies. This chapter will argue for the importance among both academics and practitioners to pull their head out from the single project perspective and understand the whole context (Sydow and Braun 2018) as it will influence construction project organising. The chapter will look beyond the individual project to inter-project relationships through empirical examples of organisational routines emerging between projects, managed by different private

Construction Project Organising, First Edition. Edited by Simon Addyman and Hedley Smyth.

developers and public agencies (i.e. construction clients), within an urban development project ecology.

By using organisational routines, as understood within routine dynamics (Feldman et al. 2021), the examples in this chapter will illustrate how routines emerge between the clients' projects in a project ecology, as coordinating mechanisms of their interdependencies. From this, the latter discussion will evolve around how projects are influenced by the emerging inter-project relationships in a project ecology, and why the inter-project perspective needs to be accounted for. To enable the discussion, the next section will address relevant project and routines literature and how it should be understood together in a construction project context. This will then be combined with the project ecology example to discuss how a project ecology and organisational routines perspective can extend the understanding of construction project organising, before ending the chapter by proposing future research avenues.

9.2 Urban Development as a Project Ecology

Urban development districts, explained as a project ecology, are created when an initiator divides an urban development area into smaller plots for different private developers and public agencies (hereinafter clients) to construct a building on it. This heterogeneous form of urban development has increased in the past decades in an effort to create viable cities, but it also calls for closer collaboration between policymakers, landowners, clients, spatial planners, and suppliers (Zakhour and Metzger 2018). Consequently, it is not only interesting to explore construction project organising from the single project perspective, but rather to also look deeper into the inter-project interdependencies as projects and the project ecology recursively influence and depend on each other, especially if we lengthen our time perspective beyond the single project and reflect on municipalities and developers programmes and sequential projects.

An urban development district, as an example of a project ecology, entails a heterogeneous mixture of actors and organisations contributing to developing a city. Urban development districts are often initiated through local government, such as a municipality owning land that they want to develop to extend or regenerate part of a city. Thereby, seen from the initiator – the municipality – perspective, urban development districts can be seen as a programme. The initiator allocates, outsources, or licences land to clients to actually develop the land into a new neighbourhood (Hedborg and Karrbom Gustavsson 2020). These clients play a central role in the development process, but, from their perspective, they focus on completing their single construction project rather than the municipality's programme. Some large clients might have their own programmes, but within the urban development district the clients only have one ongoing project. These single construction projects in the district become interdependent through their spatial closeness. In turn, the clients will procure consultants, contractors, and suppliers to plan, design, and construct according to their specifications. The majority of actors will come from a rather limited, regional market. In such contexts, actors come from different professions and organisations working for different projects, with varied project goals, all contributing to the development of a district.

9.3 Project Ecologies and Its Temporary Organising

Projects are often described through temporary organising (Bakker et al. 2016), organising to achieve pre-defined tasks by various actors within a given time frame (Bakker et al. 2016; Lundin and Söderholm 1995). However, it is not that simple; temporary organising will also be defined by the actors' previous experiences and future expectations of the other actors and organisations (Stjerne et al. 2019; Reinecke and Ansari 2015). Moreover, the actors involved often come from different sectors, organisations, and professions, adding to the complex web of interdependencies (Bechky 2006; Grabher 2004; Braun and Lampel 2020).

Project ecologies are based on the understanding that it is a context *'in which incongruent physical and organizational layers are "stapled" for a limited period of time – just to be reconfigured anew in the context of subsequent projects'* (Grabher 2002, p. 259). Following Giddens (1984) perspective of a duality between structure and agency, Grabher suggests that projects cannot be separated from their context; project and ecology are mutually constituted. More specifically, a project ecology is defined as *'the interdependencies between projects and the particular firms, personal relations, localities and corporate networks from which these projects draw essential sources'* (Grabher 2002, p. 246).

Project ecologies entail coordinating between projects, not just the formal coordinating through organisational ties and contracts but highlighting the informal coordinating (Grabher 2004). A complexity that must be handled between projects is the potential rivalry, where actors are required to collaborate and compete interchangeably (Grabher 2004; Lundin et al. 2015). Within a project ecology, actions between the actors in the different projects can informally be developed into routines, focusing on coordinating between the projects (Sydow 2009; Defillippi and Sydow 2016). In the case of urban development districts, Karrbom Gustavsson (2016) found that changes and crisis were constantly needed to be handled where rationalistic planning was impossible, and the actors had to adjust their actions to the ever-changing context. Formal routines stemming from planning, and incorporated into organisations and contracts, are common both within single construction projects (see for instance Addyman et al. 2020) and at a potential programme level (see for instance Martinsuo and Hoverfält 2018). However, between construction projects managed by different clients, there are no formal relationships at project initiation; hence, it is through informal organising that relationships are built and routines emerge, potentially becoming formalised through contracts over time.

9.4 Organisational Routines in a Project Context

Even though project studies and studies within construction management have moved beyond the notion of searching for 'one best solution', it is only lately that process perspectives have gained prevalence (Bakker et al. 2016; Vaagaasar et al. 2020). This implies that we consider the context as being dynamic (Sydow and Braun 2018), which can enable increased understanding of emerging inter-organisational relationships (Sydow and Staber 2002), by *'engaging with the everyday realities of organizational life that are rich with contingency, multiplicity, and emergence'* (Feldman and Orlikowski 2011, p. 18). For project ecologies, this means that the ecology is constantly reproduced by actors through their actions without fully controlling its structure, all the while the structure influences the actions taken.

In project contexts, actors take actions to overcome their differences and move towards mutual understanding through conflicts and trade-offs, which leads to new practices emerging such as new or recreated routines (Stjerne et al. 2019). To discuss routines in a project context is rather new (Cacciatori and Prencipe 2021; Bygballe et al. 2021), as routines have been thought of as standardised and repetitive work while projects as being ad hoc and unique (Grabher 2002). Bygballe et al. (2021) illustrate how a routine perspective is helpful in understanding change, learning, and capabilities in project organising. Repetitive actions emerging into routines will stem from both new actions in the new project contexts and from past experiences from previous projects. This perspective on routines *'is fundamentally about going "beyond routines as things" to understanding that the routines are ongoing, unfolding processes'* (Feldman et al. 2021, p. 11).

From a routine dynamics perspective, organisational routines are understood as being situated actions performed by multiple actors, influenced by artefacts, and patterned into routines over time (Feldman and Pentland 2003; Feldman 2016). Feldman et al. (2021) have concluded the following four aspects to describe routines:

- The orientation to accomplish particular tasks, even though results might differ;
- Routine enactment involves a particular sequence of actions, which can direct focus to variations;
- The actors' familiarity with a recurrent action pattern, earlier experiences provide competences and references;
- Attempts to reflective regulation, managing, and influencing through artefacts.

Due to the complex organising processes in projects, and especially project ecologies, routines to a large extent emerge as coordinating mechanisms (Parmentier-Cajaiba et al. 2021). By following routines in project ecologies, we can shed light on the repetitiveness between projects, and thereby, routines can help to shift the focus away from actions within the single projects to those between projects. These repetitive actions can for example be working together on sustainability requirements imposed by local government, setting up meetings to coordinate logistics and hiring joint resources both with parallel projects and over project life cycles. In the next sections, some illustrative examples will be used to shed light on routines emerging between construction projects.

9.5 The Case of an Urban Development District

An urban development district in the outskirts of Stockholm City will serve as an empirical example of an urban development project ecology with a high density of construction project organising. In this case, illustrative examples of two inter-project routines will be described.

9.5.1 Case Background

The case study was performed in one of Sweden's largest ongoing urban development districts, which includes the regeneration and extension of an old harbour area where the land is mainly owned by the municipality of Stockholm, which is also the initiator of the development. The development plans include around 12 000 new apartments and workspace for 35 000 people, combined with infrastructure and commercial spaces. The municipality have planned for the work to be conducted over 20 years, and it started around 2010.

The municipality divided the work into sequential stages (here called district stages), planning for one limited area at a time. The municipality allocates land to 10 or so different clients in each district stage in order to perform housing projects. Due to this set-up, several different clients will build next to each other, sharing physical space during the construction phase but must also plan for shared space during the lifetime of the completed constructions. Moreover, the municipality has, in the land allocation agreements, stipulated requirements on the clients as to how they must build large garages together connecting the buildings and blocks underground and how they must allocate public spaces between the buildings. To add further complexity for the clients, the municipality have decided on ambitious sustainability goals for the entire area, stipulating requirements which force the clients to use new techniques and products not yet tested on the market. This will affect the clients both in how they will design their buildings and in that some goals have to be coordinated on a block level, that is, with one or two other clients. All these aspects serve as fertile soil for emerging inter-project action patterns.

As the development and construction industry is largely regional, many of the consultants and contractors involved will work in the district in more than one district stage and sometimes even in several projects in one stage, in different constellations. For example, the clients having won land allocation competition in more than one district stage, contractors having been procured by several clients, or consultants specialised in a small field and thereby procured in several projects due to lack of competition. This means that both past experiences and future expectations are involved when actors perform actions within and between the projects, but also enables the actors to informally share experiences between each other.

9.5.2 A Note on Collecting and Analysing the Material

The material presented in this chapter is compiled from a case study previously presented in a number of articles (Hedborg Bengtsson et al. 2018; Hedborg et al. 2020; Hedborg and Karrbom Gustavsson 2020). The case study, following clients in the urban development, was conducted between 2016 and 2019 to get an in-depth understanding of the context (Stake 1995). The collected material includes meeting observations between clients' project managers and managers at the municipality, interviews with these actors, and document studies of agreements and informative texts by the municipality.

During the almost four years of collecting material, the focus has increasingly been placed on inter-project interdependencies, more specifically focusing on the client's interdependencies, their actions to handle the interdependencies over the project boundaries, and the influence from the municipality. This process of continuously developing the research design has been carried out to ensure reflexivity between empirical material and theory (Alvesson and Sköldberg 2009).

The material has been gathered in three sub-studies. The first sub-study concerns pre-study interviews of clients, contractors, and municipal managers in the first three stages (district stages 1–3) of the urban development district. This pre-study mainly served to get an understanding of the context and to develop the research design for the two upcoming sub-studies. The second sub-study is a longitudinal study following three clients more closely in one stage (district stage 4) from planning to completion. With this study, a deeper understanding of the clients' project manager's work was gained and more specifically what issues where prevalent for them in executing their projects and how they handled these issues. For the third sub-study, the planning phase of a sequential stage (district stage 5) was followed by observing planning meetings between the clients and the municipality and conducting follow-up interviews with all participants. This sub-study provides insights into the coordinating that the clients carried out between their projects from the start, and the emerging routines stemming from their actions.

The identified emerging inter-project routines from analysing the material will serve as an illustrative example of routines emerging between projects. Using routine dynamics and the four presented aspects of routines has enabled ascribing words and relations to these practices (Feldman et al. 2021). This enables the following discussion on better understanding inter-project interdependencies, the emerging relationships from them, and why a project ecology perspective can gain new insight into construction project organising.

9.6 Illustrative Examples of Inter-Project Routines

By the analysis of the empirical material, it is visible that clients start to build relationships from the structure of the project ecology and that these relationships are expanded over their project life cycles as more trust is gained between the clients. During interviews, the clients discuss these relationships as important for their project completion, and it is clear that the inter-project interdependencies are essential to account for when performing projects in a project ecology. As the clients' role is extending into both intra-project work and inter-project coordinating.

Seen through Feldman et al.'s (2021) four aspects characterising routines, it is possible to describe some specifics about routines emerging between projects, performing the inter-project coordinating. These specifics will be illustrated through the descriptions of two inter-project routines below. To summarise the main points, *first*, the tasks that are to be accomplished are difficult to predict upfront as each project is planned for separately with different goals. *Second*, from this, the sequence of actions often starts with a reaction to external events from the municipality or other clients. *Third*, the understanding of the recurrent nature, or familiarity, of the action pattern varies between the clients depending on past experiences and future aims. Some clients take actions to lead in order to ensure that their wishes will be accounted for, while other clients take a more back-seat approach and follow the larger, more experienced clients in order to save resources. *Fourth*, actions are not mainly performed through following artefacts; instead, artefacts such as contracts emerge through performing actions together. The findings show that informal coordinating often precedes formal coordinating.

9.6.1 The Procuring Together Routine

The design of the municipality's land agreements with the clients has led to two to three clients sharing blocks with each other. Thereby, the clients and their appointed contractors will for example have to coordinate their construction processes (such as crane location), designing and constructing together (such as their adjacent backyards), and collaborating on sustainability requirements (such as meeting the pre-established green area percentage on a block level).

During the pre-study interviews (district stages 1–3), some clients reflected back on their completed projects to ask whether their projects would have run more smoothly if they would have procured consultants and contractors together with their neighbouring clients, both to minimise the number of actors from different organisations in the stage and to simplify the inter-project coordinating. In the sequential district stage 4, two neighbouring clients jointly procured one contractor influenced by earlier reflections. During interviews, the clients' project managers said that they were inspired to do so partly because they had heard of the difficulties to coordinate in the previous stages. However, during follow-up interviews at the time of the completion of their projects, they were unsure of its success. Halfway through the construction phase, they had to add another construction manager and separate the on-site offices in order for both projects to get sufficient management on site. Following these experiences that were shared informally between local clients, the clients in district stage 5 said during the interviews that they would most likely procure their own building contractor, based on the previous difficult experiences with shared contractors and due to the clients' different business models. From informal discussions, the clients used experiences from previous stages to accomplish the procurement task.

Moreover, another issue was how to jointly procure the clients' adjacent and shared spaces of backyards and garages. From realising their need to plan upcoming construction work together, the clients started to initiate meetings more and more regularly together. From these emerging meeting fora, they decided to procure some design consultants together in their blocks. For example, in one block, they decided to have a joint fire consultant as some fire escape routes would pass through each other's properties. Another example is that in some blocks, the clients' procured project management consultants to manage and coordinate issues affecting all clients.

From these findings, patterns of actions can be identified both within one district stage and between sequential stages, from experience sharing and actors reoccurring in later stages. From the thoughts of potential benefits of joint procurement, this was tested in later procuring processes, and actions were taken by the clients based on both past experiences and expectations of how the procuring actions and decisions would affect their projects and project outcomes. In sum, a routine of joint procurement emerged over time and with variations from previous experiences and changed contexts, altering the actions in an interplay between formalising through contracts and keeping informal relationships to adjust to the dynamics for single project autonomy and a need for inter-project coordinating.

9.6.2 The Meeting Routine

By following the planning phase in district stage 5 closely, the process of developing spaces for meetings became visual. In the land allocation agreements for this stage, it was specified that two large garages were to be built under the five blocks of apartment buildings. During the initial planning meetings between the clients and the municipality, it became clear for the clients that the municipality would not involve themselves in how this would be carried out. The clients had to come up with a solution on how the garages should be designed, constructed, and maintained by themselves. This became a major issue for the clients and an influential interdependency between the projects, influencing individual project processes.

Based on these initial actions, different meeting fora took form with different regularity, to accomplish the coordinating task. One meeting forum that occurred rather frequently was between the clients sharing a block as they had most interdependencies between each other. Another forum with regular contact was between those clients sharing a garage; this forum was used to present and debate alternative solutions for their joint construction of the garages. An additional meeting forum was established between all clients in the district stage, sharing potential solutions between each other and discussing how to handle requirements from the municipality. From the experience of these three levels of meeting fora, special task groups took form to look into specific issues, one example being bringing in larger clients' in-house technical experts to investigate possibilities to move the garages to nearby caverns. A single client who was especially invested in a certain question often initiated the task groups.

Early in the planning phase sequence of actions occurred, the clients started to understand that regular contact was needed between them; different informal meeting fora then emerged from each other. Informal here illustrates that the meetings were not included in artefacts such as land allocation agreements or in the municipality's programme management, but initiated by the clients themselves without any formal agreements between them, and thus part of the emerging structure of the project ecology. Creating a meeting forum became a recurrent response when facing a new interdependency, such as coordinating ground levels or designing the shared garages. This meeting routine influences other routines emerging such as a contracting routine, when the clients started to establish contracts between each other as a way to formalise their relationships.

Through these two examples of emerging inter-project routines, it is illustrated how the clients' actions are patterned into routines as coordinating mechanisms that influence and are influenced by the project ecology. Through the routines, the clients are able to accomplish their own project tasks, while also maintaining the project ecology.

9.7 Discussion

Following from the two examples of routines, the following discussion will show how the notion of project ecologies can further our understanding of construction project organising, how inter-project routines influence this organising in project ecologies, and discuss the client's role in such a context.

9.7.1 Project Ecologies Further the Understanding of Construction Project Contexts

Previous research on construction project organising has focused much attention on the interdependencies between temporary and permanent organisations (Winch 2014), vertical relationships in other words, as permanent organisations have decision power in commencing and terminating projects – a principal–agent relationship. Horizontal interdependencies, on the other hand, have not gained much attention (Pryke et al. 2018). Horizontal interdependencies arise between actors that do not have formal decision-making power over each other, in this case between actors in neighbouring projects in a project ecology.

Coordinating between projects in a project ecology differs from coordinating other types of multi-project contexts, such as portfolios and programmes where the coordinating builds from formal relationships and front-end decisions (Martinsuo and Hoverfält 2018). In project ecologies, the coordinating instead develops from informal relationships, which might be regulated through artefacts over time such as the contracting routine emerging from the meeting routine described in the case. This can be compared to intra-organisational self-organising processes discussed by Pryke et al. (2018), balancing between informal and formal relationships.

The findings from the urban development project ecology indicate that the coordinating effort is placed on the internal project actors, such as the clients' project managers, horizontally between projects. The project managers must spend time and resources to collaboratively manage interdependencies between projects. Thereby, they will become dependent on the timelines of other projects and will have to share space. A benefit of this is the low cost of the organising, not having to spend resources on formal inter-project management, compared with programme management. At the same time, formal management follow-ups and conflict management is missing. With the emergence of informal leaders, the benefits from the coordinating actions might be skewed, for example when a larger client initiates the investigation of solutions beneficial for themselves. It will be a balancing act in preserving project freedom while formalising some processes to make the between organising beneficial.

All in all, the ecologies will largely influence single project processes, and the actions taken within single projects will influence the ecology. Going back to the role of the initiator, in the present case a municipality, how they plan the urban development district will influence the structure of the emerging project ecology. This will affect the outcome of the development and the new neighbourhood that the citizens will use. For example, if the municipality split the land and allocate it to many different clients, it can lead to vivid and interesting architecture. At the same time, many different housing cooperative and tenancy associations must organise themselves for maintaining the shared spaces, such as backyards and garages. The municipality's decisions will also influence how the clients can run their projects, and thereby, it will affect business plans and profit. An initiator of an urban development project ecology, in most cases a local government actor, must be aware of how their initial actions of planning for an urban development district will have a ripple effect in the project ecology, and thus potentially the performance and profitability of the projects for the client(s).

9.7.2 Inter-Project Routines Influence on Urban Development Project Ecologies

As we saw in the illustrative examples, clients' actions are patterned into routines, as defined by Feldman and Pentland (2003). This can be compared to engineered routines within projects, from front-end planning and re-enacting routines from artefacts across project phases (e.g. Addyman et al. 2020). The intra- and inter-project processes run in parallel; thereby, the internal project routines and the emerging inter-project routines are mutually constituted, where actors alter between prioritising their own projects and handling the surrounding context in the ecology. Using Feldman et al.'s (2021) four aspects of characteristics of routines, routines between projects can be specified through: (i) the difficulty in predicting tasks to be accomplished; (ii) sequences of actions follow from reacting to external events; (iii) clients having varied familiarity with the patterns of actions; and (iv) artefacts following from actions.

The emergence of routines in project ecologies should be seen as an iterative process, where some actions do not reoccur while others become recurrent sequence of actions. The project ecology of an urban development district is a complex and dynamic structure where actors will not be able to see the full picture of the ecology, but might instead only be familiar with part of a routine, in line with their profession or organisational belonging. Consequently, it might be difficult to re-enact routines from past experiences, which is common for internal project routines (Cacciatori and Prencipe 2021). Instead, the actors have to come together and take actions based more on future expectations. The routines emerging between projects illustrates how interdependencies will create shared practices and relationships over time. Relationships that are informal to start with and subsequently built from coordinating their interdependencies. This is mainly done without regulating through existing artefacts.

9.7.3 The Client Role in Urban Development Project Ecologies

As mentioned above, the clients in a project ecology will experience both rivalry and collegiality with their neighbouring projects, in line with Grabher's (2004) initial description of actors in a project ecology. This means that the clients must become ambidextrous in prioritising both their own project goals and the benefits for the project ecology. For example, they work together to handle interdependencies between their projects to the extent that routines emerge, but they will also compete over potential end customers who will rent or buy their completed apartments. Trust must thereby be seen as a vital aspect in upholding a project ecology. As trust takes time to build up in temporary organising (Bechky 2006; Bygballe et al. 2016), projects with long lifespans can greatly facilitate the emergence and maintenance of project ecologies. This, in turn, is essential as a single client through its single construction project cannot deliver the large goal of creating new neighbourhoods as a response to the ongoing urbanisation.

Viewing a multi-project context from the perspective of a project ecology will also enable the clients to visualise a broader network of actors influencing their project outcome. As the clients cannot focus only on their vertical relationships with procured suppliers but must realise the potential with their horizontal interdependencies. Moreover, to think in terms of emerging routines will deepen the understanding of the network that the clients work in, i.e. to not just focus on the nodes in a network (Grabher and Ibert 2011; Feldman 2016). This can enable movement beyond an input–output perspective to a more circular understanding of interdependencies and emerging relationships, including reflecting how and what the clients can learn from each other. The project ecology focus also illustrates a new side of clients' practices, in that it is not just about procuring suppliers, but more about coordinating and organising processes.

9.8 Concluding Thoughts and Further Development

This chapter has argued for a developed understanding of construction project organising by shifting focus away from the single project, to the spaces between projects. Project ecologies are presented as a valuable perspective to understand a multi-project context that cannot be fully understood through the lenses of programmes and portfolios. With the notion of project ecologies, the understanding of complex construction project contexts is broadened and suggests that some of the main complexities lay not within but between projects. For clients to be able to plan construction projects to include the space between projects, there is a need to plan for the resources needed for inter-project actions. Moreover, for academics to be able to follow processes beyond formal ties, there is a need to visualise the space between projects and the horizontal relationships emerging in this space. A deeper understanding of practices in urban development project ecologies is beneficial in the planning of upcoming urban development, which will influence how the ongoing urbanisation can be developed in line with United Nations sustainable development goals for cities and communities.

From these concluding thoughts, it could be interesting in further research on construction project organising to follow if and how other construction project actors take joint actions following interdependencies, such as designers, main contractors, and subcontractors working side by side with neighbouring projects. Moreover, to follow the learning processes in project ecologies between sequential projects can aid the understanding of informal knowledge sharing between different organisations involved in a project ecology.

References

Addyman, S., Pryke, S., and Davies, A. (2020). Re-creating organizational routines to transition through the project life cycle: a case study of the reconstruction of London's bank underground station. *Project Management Journal* 51 (5): 522–537.

Alvesson, M. and Sköldberg, K. (2009). *Reflexive Methodology: New Vistas for Qualitative Research*. London: Sage Publications.

Bakker, R.M., Defillippi, R.J., Schwab, A., and Sydow, J. (2016). Temporary organizing: promises, processes, problems. *Organization Studies* 37 (12): 1703–1719.

Bechky, B.A. (2006). Gaffers, gofers, and grips: role-based coordination in temporary organizations. *Organization Science* 17 (1): 3–21.

Blomquist, T., Hällgren, M., Nilsson, A., and Söderholm. (2010). Project-as-practice: in search of project management research that matters. *Project Management Journal* 41 (1): 5–16.

Braun, T. and Lampel, J. (2020). Introduction: tensions and paradoxes in temporary organising: mapping the field. In: *Tensions and Paradoxes in Temporary Organizing* (ed. T. Braun and J. Lampel). Emerald Publishing Limited.

Bygballe, L., Swärd, A., and Vaagaasar, A. (2016). Coordinating in construction projects and the emergence of synchronized readiness. *International Journal of Project Management* 34 (8): 1479–1492.

Bygballe, L.E., Swärd, A., and Vaagaasar, A.L. (2021). A routine dynamics lens on the stability-change dilemma in project-based organizations. *Project Management Journal* 52 (3): 278–286.

Cacciatori, E. and Prencipe, A. (2021). Project-based temporary organizing and routine dynamics. In: *Cambridge Handbook of Routines Dynamics* (ed. M. Feldman, B. Pentland, L. D'Adderio, et al.). Cambridge, UK: Cambridge University Press.

Defillippi, R. and Sydow, J. (2016). Project networks: governance choices and paradoxical tensions. *Project Management Journal* 47 (5): 6–17.

Engwall, M. (2003). No project is an island: linking projects to history and context. *Research Policy* 32 (5): 789–808.

Feldman, M.S. (2016). Routines as process: past, present, and future. In: *Organizational Routines: How they Are Created, Maintained, and Changed* (ed. J.A. Howard-Grenville, C. Rerup, A. Langley and H. Tsoukas). Oxford: Oxford University Press.

Feldman, M.S. and Orlikowski, W.J. (2011). Theorizing practice and practicing theory. *Organization Science* 22 (5): 1240–1253.

Feldman, M.S. and Pentland, B.T. (2003). Reconceptualizing organizational routines as a source of flexibility and change. *Administrative Science Quarterly* 48 (1): 94–118.

Feldman, M., Pentland, B., D'Adderio, L. et al. (2021). What is routine dynamics. In: *Cambridge Handbook of Routine Dynamics* (ed. M. Feldman, B. Pentland, L. D'Adderio, et al.). Cambridge, UK: Cambridge Univeristy Press.

Giddens, A. (1984). *The Constitution of Society: Outline of the Theory of Structuration*. Berkeley, CA: University of California Press.

Grabher, G. (2002). The project ecology of advertising: tasks, talents and teams. *Regional Studies* 36 (3): 245–262.

Grabher, G. (2004). Temporary architectures of learning: Knowledge governance in project ecologies. *Organization Studies* 25 (9): 1491–1514.

Grabher, G. and Ibert, O. (2011). Project ecologies: a contextual view on temporary organizations. In: *The Oxford Handbook of Project Management* (ed. P.W.G. Morris, J. Pinto and J. Söderlund). Oxford: Oxford University Press.

Hedborg Bengtsson, S., Karrbom Gustavsson, T., and Eriksson, P.E. (2018). Users' influence on inter-organizational innovation: mapping the receptive context. *Construction Innovation* 18 (4): 488–504.

Hedborg, S. and Karrbom Gustavsson, T. (2020). Developing a neighbourhood: exploring construction projects from a project ecology perspective. *Construction Management and Economics* 38 (10): 964–976.

Hedborg, S., Eriksson, P.-E., and Karrbom Gustavsson, T. (2020). Organisational routines in multi-project contexts: coordinating in an urban development project ecology. *International Journal of Project Management* 38 (7): 394–404.

Karrbom Gustavsson, T. (2016). Organizing to avoid project overload: the use and risks of narrowing strategies in multi-project practice. *International Journal of Project Management* 34 (1): 94–101.

Lundin, R.A. and Söderholm, A. (1995). A theory of the temporary organization. *Scandinavian Journal of Management* 11 (4): 437–455.

Lundin, R.A., Arvidsson, N., Brady, T. et al. (2015). *Managing and Working in Project Society*. Cambridge, UK: Cambridge University Press.

Martinsuo, M. and Hoverfält, P. (2018). Change program management: toward a capability for managing value-oriented, integrated multi-project change in its context. *International Journal of Project Management* 36 (1): 134–146.

Parmentier-Cajaiba, A., Lazaric, N., and Cajaiba-Santana, G. (2021). The effortful process of routines emergence: the interplay of entrepreneurial actions and artefacts. *Journal of Evolutionary Economics* 31 (1): 33–63.

Pryke, S.D., Badi, S., Almadhoob, H. et al. (2018). Self-organizing networks in complex infrastructure projects. *Project Management Journal* 49 (2): 18–41.

Reinecke, J. and Ansari, S. (2015). When times collide: temporal brokerage at the intersection of markets and developments. *Academy of Management Journal* 58 (2): 618–648.

Söderlund, J. (2004). On the broadening scope of the research on projects: a review and a model for analysis. *International Journal of Project Management* 22 (8): 655–667.

Stake, R.E. (1995). *The Art of Case Study Research*. Thousand Oaks, CA: Sage Publications.

Stjerne, I.S., Söderlund, J., and Minbaeva, D. (2019). Crossing times: temporal boundary-spanning practices in interorganizational projects. *International Journal of Project Management* 37 (2): 347–365.

Sydow, J. (2009). Path dependencies in project-based organizing: evidence from television production in Germany. *Journal of Media Business Studies* 6 (4): 123–139.

Sydow, J. and Braun, T. (2018). Projects as temporary organizations: an agenda for further theorizing the interorganizational dimension. *International Journal of Project Management* 36 (1): 4–11.

Sydow, J. and Staber, U. (2002). The institutional embeddedness of project networks: the case of content production in German television. *Regional Studies* 36 (3): 215–227.

Vaagaasar, A.L., Hernes, T., and Dille, T. (2020). The challenges of implementing temporal shifts in temporary organizations: implications of a situated temporal view. *Project Management Journal* 51 (4): 420–428.

Winch, G.M. (2014). Three domains of project organising. *International Journal of Project Management* 32 (5): 721–731.

Zakhour, S. and Metzger, J. (2018). From a "planning-led regime" to a "development-led regime" (and Back again?): the role of municipal planning in the urban governance of Stockholm. *disP – The Planning Review* 54 (4): 46–58.

10

Reflective Practices and Learning in Construction Organisations via Professional Communities of Practice

Meri Duryan

Associate Professor of Enterprise Management, Bartlett School of Sustainable Construction, University College London, UK

10.1 Introduction

Much of the literature on knowledge management (KM) in construction organisations highlights challenges to learning and knowledge transfer. This is posed by the fragmented nature of project operations, the complexity and interdependency across functional boundaries, and their one-off nature delivered by decentralised project teams (Carrillo 2004; DeFillippi 2001; Dubois and Gadde 2002; Duryan and Smyth 2019; Gann and Salter 2000; Grabher 2002; Love 2009; Ruikar et al. 2009; Smyth and Duryan 2016; Wanberg et al. 2017). A growing number of scholars have also acknowledged that 'no project is an island' (Engwall 2003), pointing to the connection of temporary projects with the relatively permanent firm level (Dalcher 2016; DeFillippi 2001; Grabher 2002; Morris 2002; Payne et al. 2019; Stjerne and Svejenova 2016). It is further evident from the literature that within many construction firms, communities of practice (CoP) have become important means for managing and transferring knowledge across project boundaries (Duryan and Smyth 2019; Love 2009; Ruikar et al. 2009; Wanberg et al. 2017). Yet, the role of CoP in making a shift from a reductionist mindset of project managers to reflective practice, and from siloed work in projects and functions to collective and collaborative reflection in- and on-action and learning at a firm or organisation level, has received very little attention.

This chapter contributes to an ongoing debate on KM and learning in large hierarchical construction organisations by mobilising the concepts of boundary spanning and reflective practice as a way of analysing the role of CoP in promoting reflective thinking and organisational learning to improve project practice. As part of a broader two-year research programme that studied the inhibitors to KM in a large infrastructure client organisation in the UK, this chapter reports upon findings that highlight the importance of programme management practices in project-based firms for overcoming the inherent contradiction between organising to meet short-term project goals and the long-term organisational learning strategy.

Construction Project Organising, First Edition. Edited by Simon Addyman and Hedley Smyth.

Although the hierarchical structure of organisations helps them improve operational efficiency through top-down coordination, hierarchically structured positions, control mechanisms, and management by objectives become obstacles to organisational learning (Argyris and Schön 1978; Mintzberg 1973). Reflective practice increases the capacity of professionals to contribute to organisational learning by encouraging them to question existing practices and norms (double-loop learning) and put deliberate efforts to interrupt habit patterns. However, it also becomes a threat to the system of rules and procedures within which professionals are expected to contribute by their technical expertise (Schön 1991).

The construction industry generally views projects as one-off, which makes them particularly prone to the 'silo' effect and adds to the complexity of learning from projects at a programme level. Professionals end up working in isolation in their functional or project silos, which works against reflection-in-action. In addition, decentralised project team working and a reductionist mindset of project leaders become obstacles to collaborative learning.

Based on the findings of this research, there is consensus in the organisation on the potential value CoP may add by crossing vertical and horizontal boundaries to encourage reflective practice, organisational learning, and decision-making at all levels of hierarchy. From the perspectives of the respondents, for CoP to add value by creating shared spaces for decision-making based on the expertise rather than the level of seniority and by enabling a transformational type of enterprise-wide organisational learning, a systemic approach to programme management, alignment of activities, and investment in programme management capabilities are required. Formal and informal collaboration mechanisms are needed at a programme level to help CoP to circumvent the barriers between the temporary and the permanent organisations. For construction project organising, the study provides a firm foundation for future work related to studying CoP in infrastructure organisations through enterprise lenses to view them as enterprise-wide forums for reflective practice and learning, rather than localised groups of practitioners.

10.2 Reflective Practice

Dewey (1933, p. 6) was the first to coin the term '*reflective practice*' as an '*active, persistent and careful consideration of any belief or supposed form of knowledge in the light of the grounds that support it*'. Practicing professionals are dealing with complex and messy problem situations for which they need knowledge and skills that go beyond textbook theories (Morris 2002). They need to experience confusion and uncertainty and be exposed to conscious criticism and change to engage in active learning and increase capacity to contribute to organisational learning (Schön 1991). Johns (1995, p. 23) defines reflective practice as the '*practitioner's ability to access, make sense of and learn through work experience to achieve more desirable, effective and satisfying work*'. Reflective practitioners can better understand the underlying causes of problems and take responsibility for their decisions and actions (Ayas and Zeniuk 2001).

Schön (1991) built on Dewey's work and examined the way in which professionals go about their daily practices. He suggested two levels of reflection: reflection-in-action and reflection-on-action. Reflection-in-action is central to the artistry of professional practice and refers to thinking while acting. It involves the observation of a situation that leads to doing things differently while the work is still underway. Reflection-on-action allows deeper thinking about a certain situation, which occurs after action and helps to make more informed decisions in future practice, using knowledge generated from previous experience (Schön 1991). To be effective and to be replicated, knowledge requires an understanding of the context in which it was generated. Given today's complex work environments, professionals need to think about what they are doing while they are doing it. Reflection-in-action allows practitioners to become 'researchers in the practice context' (Schön 1991, p. 68). It stimulates double-loop learning, a transformational type of organisational learning with the emphasis on strategic thinking (Argyris and Schön 1978).

10.2.1 Reflective Practice in Construction Organisations

Much of the valuable knowledge on project management is '*inherently not scientific*' and managers learn through practice (Morris 2002, p. 89), so to deal with the ambiguity of challenges in complex and unstable environments, employees need to develop their reflective skills (Schön 1991; Winter et al. 2006). There is agreement among academics that experienced practitioners develop their knowledge and capabilities through reflective practice and experiential learning (Crawford et al. 2006; Schön 1991). Oeij et al. (2017) link the model of the reflective practitioner to the theory of organisational learning and posit that reflective practice provides a basis for continuous professional development.

Projects are temporary organisations that are not enabled to have a memory (e.g. Dubois and Gadde 2002). Project completion means the end of collective learning (Schindler and Eppler 2003) as project leaders normally focus on highly specialised tasks to deliver projects on time and on budget rather than to work towards a common enterprise goal to learn and develop. The biggest part of lessons captured during project post-mortems mainly remains tacit, sticky (Szulanski 2000), and is rarely used in new projects. This is particularly critical for the construction industry considering that the processes in project-based organisations (PBO) are often unique and design-driven (Gann and Salter 2000), and the tacit form of knowledge is predominant (e.g. Szulanski 2000). Some of the good practices are normally assimilated and transferred on an ad hoc basis, relying upon individuals taking responsible action (Smyth and Duryan 2016).

A transformational type of organisational learning with the emphasis on reflective practice is particularly important for construction PBO considering that no project is an island (Engwall 2003) but a temporary organisation that is embedded in a permanent environment. According to Winch (2014, p. 728), project organising can be seen as a configuration of permanent PBO that form a '*temporary coalition to deliver an outcome*'. An enterprise perspective on reflective practice and learning from projects is increasingly necessary to avoid a tendency to 'reinvent the wheel' when faced with similar challenges in new projects, to reduce costly mistakes, enable cross-functional integration, and improve chances of successful project outcomes (DeFillippi 2001; Gann and Salter 2000; Grabher 2002; Morris 2002).

The term 'enterprise' represents a boundary-defining lens that calls for a systemic approach to the 'management and alignment of activities' (Purchase et al. 2011, p. 19) and is used in this chapter to depict interdependencies and interrelationships across boundaries of organisations involved in construction projects. Nightingale (2002, p. 1) defines enterprises as '*complex, highly integrated systems comprised of processes, organisations, information and supporting technologies, with multifaceted interdependencies and interrelationships across their boundaries*'. An enterprise perspective gives a better understanding of an organisation as a complex whole rather than combination of parts. Construction enterprises need project leaders and practitioners who can go beyond the execution-focused efficiency of project management and develop capabilities to think, reflect, and act at the level of enterprise-wide project management (Crawford et al. 2006; Dalcher 2016), which also focuses on the effectiveness of projects (Morris 2002).

Both reflection-on-practice and reflection-in-practice are essentially individual and asocial (Schön 1991). To develop capabilities to deal with ill-defined complex problem situations, organisations need to facilitate a context-specific learning environment that encourages social interaction, prioritises spaces for reflection, and facilitates group simulations and double-loop learning (Crawford et al. 2006; Lave and Wenger 1991). This may be achieved through CoP where employees increase learning through collective reflection on specific issues that concern their professional practice.

10.3 Communities of Practice as Boundary Spanners in Construction Firms

The term 'community of practice', or CoP, was first introduced by Lave and Wenger (1991) and is a social learning theory. Brown and Duguid (1991) describe CoP as a group of people who work together to share individual practices and perceptions. The concept proved to have high potential in crossing horizontal and vertical boundaries (Love 2009; Ruikar et al. 2009; Wanberg et al. 2017; Wenger 2010). A 'boundary' is a borderline that demarcates the limits of an area that may include social, cognitive, hierarchical, geographical, and relational knowledge with cultural and other divisional and disciplinary boundaries (Carlile 2002).

Large infrastructure organisations, hierarchical bureaucracies with vertically organised functions and project teams, are known for internal boundaries between projects, processes, stakeholders, and functions (Gustavsson and Gohary 2012) that inhibit cross-functional and inter-project cooperation and learning (e.g. Duryan and Smyth 2019). There is also a strongly held belief among academics and practitioners that the uniqueness and temporary nature of each project is a major obstacle for learning from projects. Some researchers, however, posit that instead of taking a position of uniqueness and temporary nature, organisations need to stress the common elements of projects and provide incentives and processes to encourage teams to share knowledge (Stjerne and Svejenova 2016; Payne et al. 2019). Projects need to be connected with the relatively permanent firm level via coordination at programme or portfolio level (Dalcher 2016; Morris 2002).

Robust KM systems and processes can support the transfer of explicit knowledge that is documented and stored in organisational repositories. However, to transfer tacit knowledge, that is intangible and resides in minds of employees, a guided joint social interaction is required at a programme or portfolio level. There is common agreement among researchers on the value of CoP for social interaction and collaborative learning in construction organisations (e.g. Love 2009; Ruikar et al. 2009; Wanberg et al. 2017), but still little is known about how CoP can overcome inherent contradiction between organising to meet short-term project goals and the long-term organisational learning and continuous professional development strategy (e.g. Dubois and Gadde 2002). It is a challenge for an enterprise to enable CoP to cross functional and inter-project boundaries to encourage collaborative learning across supply chains given the decentralised nature of project work, which adds to horizontal and vertical differentiation.

10.3.1 Reflective Practice in Communities of Practice

Previous research has focused on lessons learnt in the construction industry, continuously highlighting the problems with capturing and reusing good practice across projects (DeFillippi 2001; Duryan and Smyth 2019; Grabher 2002). Project governance in infrastructure organisations is driven from the top and has predefined rules and routines that become obstacles to inter-project learning and realisation of business benefits (Thiry 2004). Hierarchically structured positions and management by objectives discourage individuals' desire to learn beyond the defined job responsibilities and transfer knowledge (Argyris and Schön 1978; Mintzberg 1973).

Coordination at an enterprise level may not be efficient if professionals work in project and functional silos without opportunities to communicate their experiences and insights to test them against the views of their peers (Schön 1991). CoP can overcome this obstacle by using their network positioning to create an informal web of knowledge brokers, and facilitate inter-project and cross-functional discussions and enterprise-wide knowledge flows.

10.4 Methodology and Methods

This study is part of a broader two-year research programme aimed at studying the inhibitors to KM in a large infrastructure client firm. The firm operates in a multi-organisational environment in the UK and relies on an extensive supply chain to deliver infrastructure programmes and projects. During the last decade, the firm has implemented various KM projects to retain and transfer knowledge across functions, regions, programmes, and projects, yet failed to gain traction to improve business performance. Tacit knowledge generated within projects was buried in reports on lessons learnt or lost because people were moved on to other projects or left the organisation. The reasons were both vertical and horizontal and included: (i) lack of perceived value of lessons learnt for successful project outcomes by the senior management resulting in low investment in developing capabilities in programme management; (ii) reliance upon individuals to capture and transfer knowledge across functional and project boundaries internally and across the supply chain, which was inconsistent and unsystematic as they were not incentivised to do so; and (iii) overreliance on information technology platforms and failure to see them as facilitators, rather than KM systems that can transfer context-specific knowledge. The lack of KM at a programme level led to systemic failures to transfer and reuse knowledge internally and externally.

This research mobilises the concepts of boundary spanning and reflective practice to understand key challenges of current KM practices and feasibility of cultivating CoP that can promote reflective thinking and organisational learning in the client infrastructure organisation. An interpretative methodology was employed to recognise the importance of human worldviews and perceptions, while preserving the systematic approach to analysis. The research approach conforms to Yin's (2003, p. 23) case study method, which involves '*empirical investigation of a contemporary phenomenon within its real-life context*' and where multiple sources of evidence have been used. The data were obtained through engaged research, which allowed the researchers to look at and examine the findings in a real-life context.

A cognitive mapping technique, used in this research to analyse the interviews, is a problem structuring method of soft operations research (Eden and Ackermann 2018). A significant strength of the tool is that it supports the elicitation of mental models and the generation of creative ideas using the participants' language. The formal basis for cognitive maps derives from Kelly's (1955) personal construct theory, which proposes an understanding of how people 'make sense' of their world by seeking to manage and control it (Eden 1989). The maps demonstrate inconsistency or contradiction in what individuals say, which can be the reason for a perceived complexity of a problem situation. The links between the nodes represent logical implications between the concepts.

Data for this study were collected through 21 semi-structured interviews with the senior management, heads of functions, and the management team of a multi-billion-pound infrastructure programme (Table 10.1). The interview transcripts, interviewers' notes, and organisational reports were examined for the presence of information about organisational practices and routines related to KM, reflective practice, and organisational learning.

Table 10.1 Interview schedule.

Senior management	Programme management director (1) Head of procurement (1) Infrastructure programme director (1) Commercial projects director (1) Programme commercial manager (2)
Infrastructure programme management	Programme director (1) Project manager (6) Discipline manager (1) Principal design engineer (1) Programme manager (2) Procurement manager (1) Programme engineering manager (1) Planning and scheduling manager (1) Commercial manager (1)
Total	21

Each interview lasted about 1.5 hours and was recorded and transcribed verbatim. The transcripts were inductively coded and translated into the cognitive maps to depict interviewees' perception of the prevailing situation. Individual cognitive maps were validated and merged into a single map to develop a unified view of multiple perspectives that was analysed using head, centrality, domain, and cluster analyses (Eden 1989). Software tools were used to deal with the complexity of the map (e.g. Brightman 2002) and identify the goals and key strategic directions as perceived by the respondents. Given the complexity of the combined cognitive map, a more simplified schematic map is presented (Figure 10.1) to demonstrate how the analysis and relevance of the findings were derived. Some clusters and nodes that add little value to the focus of this chapter are hidden.

10.5 Findings and Discussion

The analysis of the map identified emerging issues that from the perspectives of the respondents require resolution to 'improve organisational safety and performance', which was highlighted by the interviewees as the organisational goal. Based on the centrality and domain analyses of the map, the emergent issues in descending order of importance are: '*organise professional CoP*', '*break professional silos*', '*become a learning organisation*', '*promote excellence in the company internally and externally*', '*improve collaboration with supply chain*', and '*create spaces for knowledge exchange and reflection-on-practice*' (underlined concepts, Figure 10.1). They are the heads of clusters, groups of concepts that are linked together and cover a specific area of the issue. These clusters, although inter-related, are separable from other parts of the map and can become the subjects for a deeper elaboration. The concept '*organise professional CoP*' has the highest domain and centrality scores, which emphasises the perceived importance of cultivation of professional CoP to achieve improvement in safety and performance.

10.5.1 Professional Communities of Practice

The cluster '*organise professional CoP*' directly contributes to the head of the map '*improve safety and performance*' through the development of organisational capabilities, collaboration across hierarchical boundaries, and contribution to long-term thinking (Figure 10.1). From the perspective of the respondents, CoP can help the organisation to draw upon the accumulated knowledge and experience by enabling collaboration and learning across hierarchical boundaries. At the same time, concerns were raised on the ability of CoP to deliver value considering inherent rigidities of hierarchical infrastructure organisations (i.e. hierarchically structured positions, project silos, scarce dialogue between the functional units, authoritarian leadership, etc.).

As the programme director mentioned:

> There is a lot of hierarchical bias in here. Depending on where you sit your opinion can be more or less important.

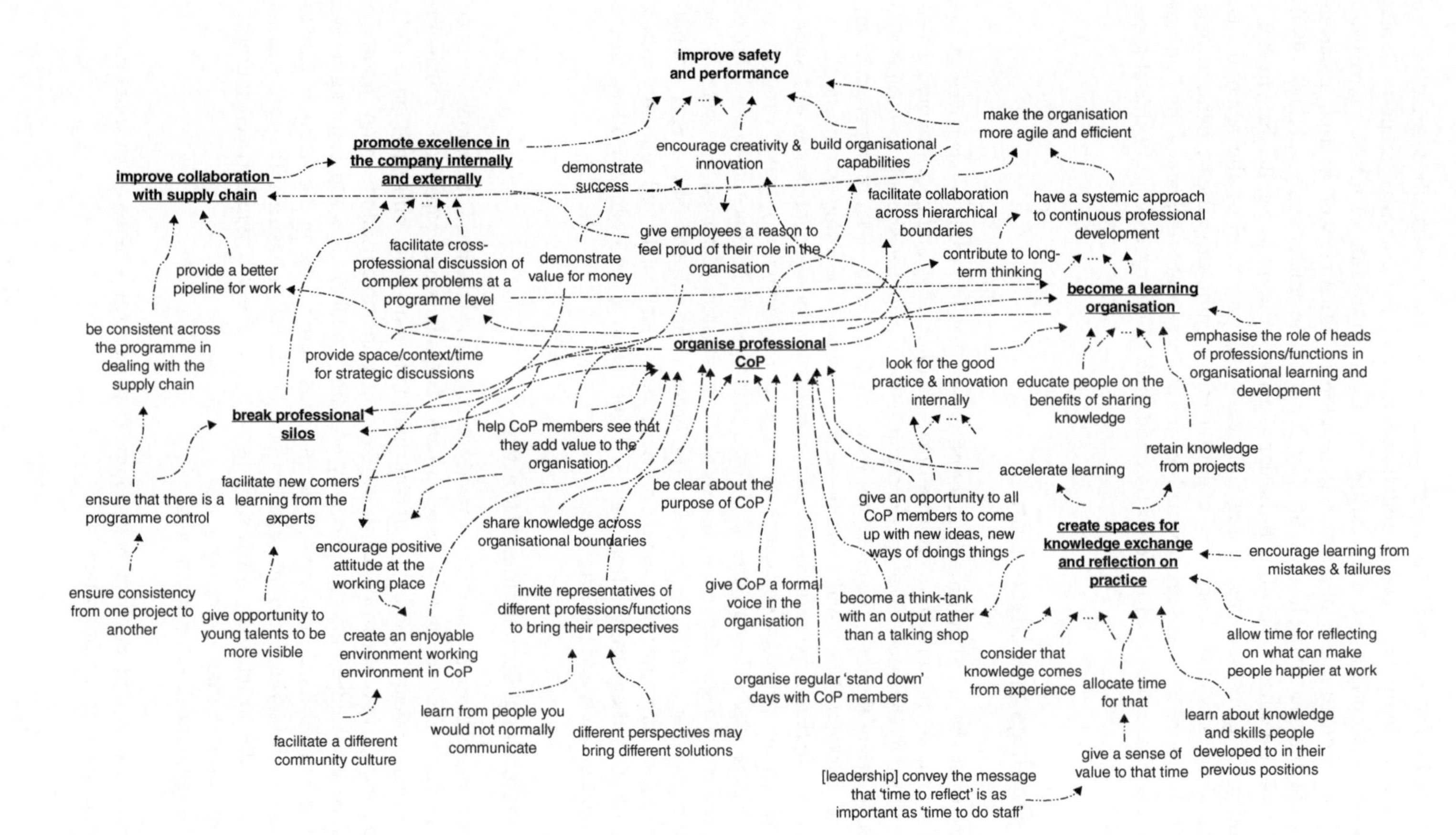

Figure 10.1 Schematic cognitive map (ellipses between arrows indicate hidden nodes).

According to a programme commercial manager:

> Bureaucracy has always been here. . . With bands we label people. Maybe it will be good to eliminate them, but we are so institutionalised; I do not think that it is possible.

There was common agreement on the shortcomings of hierarchical bureaucracy that can influence CoP. The respondents believe, however, that CoP can escape the rules imposed by the bureaucratic system (Duryan and Smyth 2019; Lave and Wenger 1991), promote social interaction across different boundaries, and enable engagement in collective learning if they are given a formal voice in the organisation (Figure 10.1).

Although collaboration and collective learning were perceived to be vital by the respondents, project managers mentioned that people try to do their best to meet specific objectives in their project and functional silos without looking across the business to see how they can contribute towards the collective organisational intelligence. As a programme commercial manager mentioned:

> I have been looking how all the functions operate; everybody is doing things in completely different ways. There is no sharing of best practice, and people are in silos. It is very confusing for stakeholders.

Purchase et al. (2011, p. 20) posits that the '*reductionist mindset has been common amongst managers within large, single organisational enterprises*', and coordination toward common 'enterprise goals' is needed if organisations want to develop core competencies to achieve performance improvement and cost reduction. The findings of this research confirm the need for programme management to link the client to the projects to enhance learning across boundaries. The head of procurement believes that CoP can increase the level of commitment of people by providing the necessary resources and linking their activities to the organisational strategic goals and objectives.

As the map demonstrates (Figure 10.1), '*organise professional CoP*' cluster contributes to the clusters '*break professional silos*' and '*promote excellence in the company internally and externally*' via the nodes '*facilitate cross-professional discussion of complex problems at a programme level*' and '*become a learning organisation*'. This implies that the respondents believe that CoP members can establish connections not only between professional communities in the same function but also between different practices given that there are relevant mechanisms at a programme level. They can contribute to organisational growth and resilience by introducing the elements of one practice into another (Lave and Wenger 1991).

The head of procurement mentioned that inviting members of different functions from the supply chain will not only bring different perspectives into discussions on specific challenges, but will also help CoP avoid becoming a 'silo thinker'. This is especially important for construction organisations as they are known for rigid boundaries between processes and functions that create silos with their attendant silo mentalities (Duryan and Smyth 2019). The facilitation of cognitive diversity in CoP will encourage outward-looking behaviours. It will affect decision-making, particularly on a client side, by providing room for flexible responses from suppliers, which, in turn, will contribute to promoting excellence in the company internally and externally (Figure 10.1).

Educating people on the benefits of retaining and sharing knowledge from projects and looking for the good practice and innovation internally are seen by respondents as the key factors that can contribute to developing a learning organisation. This is what a programme manager thinks about the disadvantages of failing to share knowledge in the organisation:

> The worst thing in this company is that you can see the groups of people doing the same thing and still not sharing their experiences; they work at cross-purposes. Knowledge sharing does not exist and when people leave the organisation, they take that knowledge with them.

Project managers, engineers, and procurement and commercial experts do not collaborate as they should. The same programme manager described their communication as '*who shouts louder gets his way in*'. A discipline manager also added:

> I do not think we really know when we need to share. In our small team we are doing better, but in a wider team we do not share. Some people do not understand what knowledge transfer is. . . We do not retain knowledge. . . we consider knowledge as a competitive advantage.

Based on the feedback by the respondents, the main challenge is to provide an environment that is conducive to learning from current and previous mistakes. Formal and informal collaboration mechanisms are required at a programme level.

CoP was perceived also as a potential forum for learning, new knowledge creation, and innovation, which are the characteristics of a learning organisation. However, the respondents posit that the organisations need to put in more effort to capitalise on tacit knowledge that is generated due to the richness of experience of people who join projects from different firms. According to the programme director:

> . . . we have got a lot of answers, lots of smart people but we do not want to know where they come from; there is no interest in their experience and knowledge. You are defined by your title. And then, we spend ridiculous amount of money on consultants. . .

Projects depend on knowledge, expertise, and 'contingent' capabilities of the organisations (Winter et al. 2006) that are mainly generated outside their organisational boundaries, and failure to capitalise on them is quite costly. The respondents believe that regular face-to-face collaboration among CoP members will contribute to shared understanding of the benefits of accumulated knowledge of their peers.

The senior management of the client organisation realises that the success of the multi-billion-pound construction programme of projects depends on the ability to create an '*effective pipeline of work*' to improve collaboration with their complex and fragmented supply chain. A programme commercial manager mentioned that their function needs to work with the supply chain to:

> . . . understand the labour, material costs, the methodology, how they put things together, their overheads and profit margins, their preliminaries. . . to build up cost plans. We all need to be working together collaboratively and communities of practice can help.

There is a shared understanding among the respondents that project and programme managers and functional leads are dealing with problems that are part of complex systems of problems. Those systems of problems interact with one another across multiple organisations involved in projects and cannot be solved in isolation; thus, a systemic approach to the management and alignment of activities is needed. This is complicated considering the hierarchical structure of the organisation with vertically organised functions and project teams. However, CoP can facilitate a risk-free learning environment and, to some extent, ensure the alignment of activities and even escape the rules, when necessary, without conflict with administrative hierarchy (Duryan and Smyth 2019).

According to a programme manager, the organisation needs to be consistent across the programme in dealing with the supply chain:

> As a business we just keep sending [to the contractors] the piles of files to read. We have processes and procedures that we do not understand ourselves, we write documents we do not understand. . . but we expect our contractors understand them.

He mentioned that CoP can help facilitate collective collaboration and learning across the supply chain to avoid repeating the same costly mistakes. Given that project managers' reflection-in-action can be limited by the learning system of the organisation (Schön 1991), they need to communicate their ideas and thoughts with other professionals across the supply chain.

10.5.2 Reflective Practice

There was agreement among the respondents that employees need time to thoroughly elaborate on their practices from different perspectives to enhance learning from both good practice and failure. According to Morris (2002), the effective management of projects requires knowledge and skill that cannot be fully covered by the relevant bodies of knowledge as it is impossible to capture all nuances of context-specific project work, especially considering that decentralised project team working in the industry is a norm. Besides, practitioners are often '*in conflicts of values, goals, purposes, and interests*' (Schön 1991, p. 17) and deal with problems in isolation and with a reductionist mindset. To keep means and ends together and to connect thinking with doing when faced with uncertainties, project leaders need to have an opportunity to collectively elaborate on insights from reflection.

As the map demonstrates (Figure 10.1), the cluster '*create spaces for knowledge exchange and reflection-on-practice*' contributes to the cluster '*organise professional CoP*' through accelerated learning. It also contributes to the cluster '*become a learning organisation*' through the retention of knowledge from projects. Previous research that elaborates on the importance of learning from construction projects continuously highlights the challenges caused by the nature of the industry (DeFillippi 2001; Grabher 2002; Love 2009; Ruikar et al. 2009; Wanberg et al. 2017). Based on the findings of this research, creating a suitable work environment and allocating time for reflection at a level of enterprise may be the key to helping construction practitioners develop a positive mindset towards achieving the goals of the enterprise.

From the perspectives of the infrastructure programme director, to encourage positive attitudes to reflection, learning, and innovation at the workplace, it is necessary to demonstrate to people that they add value to the organisation (Figure 10.1). He posits that unhappy people at the office 'become institutionalised' and normally stop challenging processes and procedures. They gradually become less able to reflect in- and on-practice and act independently as they get used to the same routines under the same rules and for a long time. This hinders reflection and double-loop learning, which are necessary for the development of capabilities to deal with ill-defined complex problem situations (Crawford et al. 2006; Lave and Wenger 1991).

A commercial manager mentioned about the barriers that are caused by the line managers who do not give importance to allocating time and space for reflection-on-action. As the commercial projects director mentioned:

> . . .if we were to walk past someone in the office who is looking out of the window and if you ask them 'what are you doing' and if they said 'thinking', we would say 'go and do something'.

He believes that the leadership needs to convey the message that '*time to reflect is as important as time to do stuff*'. Professional CoP, where the only real status is that of a member, can be a place to overcome artificial barriers created by some managers and help professionals express themselves and think on the improvement of current practices. By understanding the importance of reflective practice and allowing themselves to experience confusion and uncertainty, managers may increase their capacity to contribute to significant organisational learning (Schön 1991).

The same commercial projects director highlighted the importance of CoP members reflecting in- and on-action to ensure that they do not become talking shops but rather think tanks with specific outputs that can be replicated internally and across the supply chain.

Reflective practice is widely considered to be an important activity for the development of capabilities and continuous organisational learning. There is a strong link between the organisational capability to learn and innovate and its ability to capitalise on the knowledge and expertise of its employees (Anumba et al. 2005; Newell et al. 2009; Sergeeva and Duryan 2021). This is particularly relevant to the construction industry and construction project organising given that project team members bring expertise and innovative problem-solving from other firms to deliver their own technical and social capabilities.

10.6 Conclusions

Given the entire scope of the transformation occurring in the industry nowadays, many construction firms and organisations are realising that the effective management of projects requires knowledge and skills that cannot be fully covered by the relevant bodies of knowledge. In addition, the mechanistic model for organising with its command-and-control systems is no longer adequate. Hierarchical boundaries in infrastructure organisations, coupled by traditional construction project practice, have made cross-functional and inter-project cooperation, integration, and learning a major challenge. To address this challenge, boundary spanning is needed to break down horizontal and vertical silos and cultivate a regular habit of reflective practice across vertical and horizontal boundaries.

A growing attention in the literature has been given to the role of CoP in facilitating learning and innovation in construction organisations; however, little attention has been paid to the value they can add by making a shift in mindsets from relying on fixed standards in predefined contexts towards a more reflective approach to allow practitioners to go beyond execution-focused project management and act at the level of programme management.

The central message that has emerged from the analysis is that the concept of CoP has not been effectively embraced by the industry as a strategic tool for boundary spanning across both vertical and horizontal boundaries to encourage reflective practice, collaborative enterprise-wide learning, and knowledge transfer to improve performance. There are several reasons for that revealed by the literature and confirmed by this research including organisation-wide disconnect between the project management and the functional and administrative groups, hierarchically structured positions, project and functional silos, decentralised ways of working, reductionist mindsets, and dominance of the local expertise of project managers. Each reason points to the lack of KM mechanisms at a programme and/or portfolio levels.

It emerged from the findings of this research that in the absence of formal and structured KM systems and given low investment levels in new capabilities and competencies at a programme level, CoP can be the key means for facilitating cross-functional and inter-project learning and reflection in- and on-action. CoP members can be seen as agents of an organisation's reflective conversation with its situation who span horizontal and vertical boundaries to facilitate organisational learning at a programme level to help construction practitioners develop a positive mindset towards achieving the goals of an enterprise. To avoid becoming a 'silo thinker' and to contribute to decision-making, particularly on a client side, a fundamental shift in the mindsets of senior management and facilitation of 'learning how to learn' (or triple loop learning) is necessary.

This chapter contributes to thinking about organisational learning and KM for construction project organising by raising the following questions:

1) How to maximise the value CoP can deliver considering rigidities of hierarchical infrastructure organisations and inherent contradiction between organising to meet short-term project goals and the long-term learning and continuous professional development strategy (specifically in a client and tier 1 supply chain)?

2) Vertical boundaries in hierarchical organisations demarcate different levels of control across functions, while horizontal boundaries bring into focus specific expertise within a function, division, project. Both types of boundaries are socially constructed distinctions that add value to organisations and, at the same time, create challenges for collaboration and learning. What skills, abilities, and knowledge are necessary for CoP members to deal with tensions and ambiguities stemming from the complexity of both types of boundaries?
3) Managers reflect on the strategies by which they can lead the organisation or enterprise to success, but neither senior managers nor middle level managers reflect on creating the conditions to 'win the game'. How to avoid situations where leaders deal with problems in isolation with a reductionist mindset and seek to gain control over the situation without considering the interests of an enterprise?
4) What will happen in large infrastructure organisations when engineers, procurement specialists, project leaders, or operatives on a construction site begin thinking and acting not as technical experts but as reflective practitioners? Will their reflection-in-action pose a potential threat to the dynamically conservative system?

Reflection is not always easy; the main challenge for any organisation, particularly for construction PBO, still remains the cultivation of a culture that will enable reflection in- and on-action. Schön (1991) draws an analogy between reflection-in-action and playing jazz. All musicians know the basic theme; at the same time, they play their parts listening to one another to align their play and collectively create music gradually evolving their ways of doing it. Being a part of a group of peers in CoP will help professionals develop the habit of engaging in the reflective process to learn from diverse opinions and perspectives and create new knowledge. CoP members will have enough flexibility to decouple from the complex network of relations in hierarchical bureaucracies to create their own culture. This can be one of the means of overcoming the inherent contradiction between organising to meet short-term project goals and long-term learning and continuous professional development.

References

Anumba, C.J., Egbu, C., and Carillo, P. (2005). *Knowledge Management in Construction*. Oxford: Blackwell Publishing.

Argyris, C. and Schön, D.A. (1978). *Organisational Learning: A Theory of Action Perspective*. Reading, MA, USA: Addison-Wesley.

Ayas, K. and Zeniuk, N. (2001). Project based learning: building communities of reflective practitioners. *Management Learning* 32 (1): 61–76.

Brightman, J. (2002). *An Introduction to Decision Explorer*. London: Banxia Software Ltd.

Brown, J. and Duguid, P. (1991). Organisational learning and communities of practice: towards a united view of working, learning, and innovation. *Organisation Science* 2 (1): 40–57.

Carlile, P.R. (2002). A pragmatic view of knowledge and boundaries: boundary objects in new product development. *Organisation Science* 13 (4): 442–445.

Carrillo, P.M. (2004). Managing knowledge: lessons from the oil and gas sector. *Construction Management and Economics* 22 (6): 631–642.

Crawford, L., Morris, P., Thomas, J., and Winter, M. (2006). Practitioner development: from trained technicians to reflective practitioners. *International Journal of Project Management* 24 (8): 722–733.

Dalcher, D. (2016). Rethinking project practice: emerging insights from a series of books for practitioners. *International Journal of Managing Projects in Business* 9 (4): 798–821.

DeFillippi, R.J. (2001). Project-based learning, reflective practices and learning outcomes. *Management Learning* 32 (1): 5–10.

Dewey, J. (1933). *How we Think*. Boston, MA: DC Heath & Co.

Dubois, A. and Gadde, L.E. (2002). The construction industry as a loosely coupled system: implications for productivity and innovation. *Construction Management and Economics* 20 (7): 621–631.

Duryan, M. and Smyth, H. (2019). Cultivating sustainable communities of practice within hierarchical bureaucracies: the crucial role of an executive sponsorship. *International Journal of Managing Projects in Business* 12 (2): 400–422.

Eden, C. (1989). Using cognitive mapping for Strategic Options Development and Analysis (SODA). In: *Rational Analysis for a Problematic World* (ed. J. Rosenhead), 21–42. Chichester: Wiley.

Eden, C. and Ackermann, F. (2018). Theory into practice, practice to theory: action research in method development. *European Journal of Operational Research* 271 (3): 1145–1155.

Engwall, M. (2003). No project is an island: linking projects to history and context. *Research Policy* 32 (5): 789–808.

Gann, D.M. and Salter, A.J. (2000). Innovation in project-based, service-enhanced firms: the construction of complex products and systems. *Research Policy* 29 (7/8): 955–972.

Grabher, G. (2002). Cool projects, boring institutions: temporary collaboration in social context. *Regional Studies* 36 (3): 205–214.

Gustavsson, T.K. and Gohary, H. (2012). Boundary action in construction projects: new collaborative project practices. *International Journal of Managing Projects in Business* 5 (3): 364–376.

Johns, C. (1995). The value of reflective practice for nursing. *Journal of Clinical Nursing* 4: 23–40.

Kelly, G.A. (1955). *The Psychology of Personal Constructs*, vol. 2. New York, N: Norton.

Lave, J. and Wenger, E. (1991). *Situated Learning: Legitimate Peripheral Participation*. Cambridge: Cambridge University Press.

Love, P.E.D. (2009). Editorial: communities and champions of practice: catalysts for learning and knowing. *Construction Innovation* 9 (4): 365–371.

Mintzberg, H. (1973). *The Nature of Managerial Work*. New York: Harper Row.

Morris, P.W. (2002). Science, objective knowledge and the theory of project management. *Civil Engineering* 150 (2): 82–90.

Newell, S., Robertson, M., Scarbrough, H., and Swan, J. (2009). *Managing Knowledge Work and Innovation*. London: Pelgrave Macmillan.

Nightingale, D. J. (2002). Lean enterprises – A Systems Perspective. *MIT Engineering Systems Division Internal Symposium*. Cambridge, MA, USA.

Oeij, P.R.A., Gaspersz, J.B.R., van Vuuren, T., and Dhondt, S. (2017). Leadership in innovation projects: an illustration of the reflective practitioner and the relation to organisational learning. *Journal of Innovation and Entrepreneurship* 6 (1): 1–20.

Payne, J., Roden, E.J., and Simister, S. (2019). *Managing Knowledge in Project Environments*. Abingdon: Routledge.

Purchase, V., Parry, G., Valerdi, R. et al. (2011). Enterprise transformation: why are we interested, what is it, and what are the challenges? *Journal of Enterprise Transformation* 1 (1): 14–33.

Ruikar, K., Koskela, L., and Sexton, M. (2009). Communities of practice in construction case study organisations: questions and insights. *Construction Innovation* 9 (4): 434–448.

Schindler, M. and Eppler, M.J. (2003). Harvesting project knowledge: a review of project learning methods and success factors. *International Journal of Project Management* 21 (3): 219–228.

Schön, D.A. (1991). *The Reflective Practitioner: How Professionals Think in Action*. Aldershot: Ashgate Publishing Ltd.

Sergeeva, N. and Duryan, M. (2021). Reflecting on knowledge management as an enabler of innovation in project-based construction firms. *Construction Innovation* 21 (4): 934–950.

Smyth, H.J. and Duryan, M. (2016). Knowledge application in the supply network of infrastructure programme management. *Proceedings of RICS COBRA*. Toronto, Canada (22–24 September 2016). London, UK: Royal Institution of Chartered Surveyors.

Stjerne, I.S. and Svejenova, S. (2016). Connecting temporary and permanent organising: tensions and boundary work in sequential film projects. *Organisation Studies* 37 (12): 1771–1792.

Szulanski, G. (2000). The process of knowledge transfer: a diachronic analysis of stickiness. *Organisational Behavior and Human Decision Processes* 82 (1): 9–27.

Thiry, M. (2004). Programme management: a strategic decision management process. In: *The Wiley Guide to Managing Projects* (ed. P.W.G. Morris and J.K. Pinto), 12. New York: Wiley.

Wanberg, J., Javernick-Will, A., and Taylor, J.E. (2017). Mechanisms to initiate knowledge-sharing connections in communities of practice. *Journal of Construction Engineering and Management* 143 (11): 1–10.

Wenger, E. (2010). Communities of practice and social learning systems: the career of a concept. In: *Social Learning Systems and Communities of Practice* (ed. C. Blackmore), 179–198. London: Springer.

Winch, G.M. (2014). Three domains of project organising. *International Journal of Project Management* 32 (5): 721–731.

Winter, M., Smith, C., Morris, P., and Cicmil, S. (2006). Directions for future research in project management: the main findings of a UK government-funded research network. *International Journal of Project Management* 24 (8): 638–649.

Yin, R.K. (2003). *Applications of Case Study Research*. Thousand Oaks: Sage.

Part IV

Inside the Project

The chapters in this part address organising inside the project.

First, Aaltonen and Turkulä inen challenge some taken-for-granted thinking in research and practice about co-location as a positive feature of and for developing collaborative practices. Their analysis of the dark side of co-location demonstrates how such organising can result in dysfunctions and pose constraints for a project.

In the second of four chapters in this final part, Ahola examines the task level of construction project management, exploring the extent of synchronisation of activities. He draws out the complexity and challenges of alignment and synchronisation in progressing successfully towards task, hence project completion.

Almadhoob builds on the work of a forerunner of network analysis in analysing projects, namely the late Stephen Pryke to whom this book is dedicated. Applying social network analysis as both a set of concepts and methods applied to research, she examines how project actors build upon and further develop networks that go beyond the hierarchical thinking and structures that are often seen as the means to rationalise action to present a different way of understanding organising.

Finally, Addyman takes a look at one function, namely procurement, to explore how the role of dialogue can be fostered and harnessed to improve collaboration and collaborative practices to transition and potentially transform project procurement and project capabilities more generally. The content of this chapter links back to the book title, the overarching theme, and to the Editorial Chapter with which the book starts.

11

The Use of Collaborative Space and Socialisation Tensions in Inter-Organisational Construction Projects

Kirsi Aaltonen[1] *and Virpi Turkulainen*[2]

[1] *University of Oulu, Industrial Engineering and Management, Oulu, Finland*
[2] *Haaga-Helia University of Applied Sciences, Helsinki, Finland*

11.1 Introduction

Inter-organisational construction projects face the challenge of aligning objectives, facilitating trust, and integrating the resources and knowledge of various autonomous organisations and project team members with differing backgrounds and objectives in a timely manner so that the planned project outcomes are realised (DeFillippi and Sydow 2016; van Marrewijk et al. 2016). Physical co-location of project team members in a shared workspace is suggested as an important mechanism for managing the organisational boundaries and the typical challenges of disintegration in construction project settings (Baiden et al. 2006). Recently, the use of co-locational spaces in construction projects has increased significantly; the so-called Big Rooms fostering the team's collaborative behaviour has been promoted as an efficient means to facilitate cross-disciplinary work, interaction, and integration of diverse knowledge among professionals, particularly in integrated project deliveries and alliances (Kokkonen and Vaagaasar 2018; Walker and Lloyd-Walker 2015). Research on collaborative spaces has primarily focused on describing and assessing the positive outcomes of co-location, for example, on socialisation, referring to the interaction and communication between different organisations. Socialisation facilitates the development of personal familiarity, improved communication, and problem-solving (Gupta and Govindarajan 2000; Van Maanen and Schein 1979) and is considered fundamental for the performance of and value creation in project teams (Aaltonen and Turkulainen 2018).

In reality, the relationship between space and the actual collaborative behaviour of individuals and socialisation is a highly complex and dynamic issue, filled with tensions and struggles and situated in space and time. Consequently, the co-location of project personnel representing different organisations in the same physical space does not automatically produce positive outcomes (Bresnen 2007). Rather, managerial practices play an important role in realising the benefits of collaborative spaces and facilitating

Construction Project Organising, First Edition. Edited by Simon Addyman and Hedley Smyth.

favourable social processes (Bosch-Sijtsema and Tjell 2017). Moreover, conflict and cooperation have been suggested to co-exist among actors working in the same collaborative space of an inter-organisational project (van Marrewijk et al. 2016), implying that although physical co-location in a collaborative space may bring significant benefits, it may also pose struggles and tensions in construction projects.

Organising in inter-organisational construction projects can be viewed as a problem-solving process (Weick 1974), where the universal challenges of task division, task allocation, provision of reward, and provision of information are continuously addressed throughout the project lifecycle (Lehtinen and Aaltonen 2020; Puranam et al. 2014). Socialisation processes and the management of socialisation tensions taking place in the collaborative space are a significant part of project organising in this context: the development of interaction and communication among project participants during a project has impact on how roles and responsibilities are shared within the team, how motivated project actors are to work toward the common goals, and how actively actors share information with each other so that they can execute tasks in a timely fashion (Puranam 2017). Understanding socialisation dynamics and tensions in collaborative spaces is, therefore, crucial for managing inter-organisational project organising, characterised by the contradictory requirements of the project actors that need to be mastered. This view on organising, therefore, also implies a shift from the linear and static logic of designing project organisations to focusing on non-linear organising processes that allow for capturing and conceptualising contradictions and countervailing processes (Schreyögg and Sydow 2010).

In practice, developing socialisation may be a more complex process than understood in research. In this chapter, we are interested in tensions that may be produced by the use of collaborative space during socialisation processes in inter-organisational construction projects. To demonstrate tensions in the socialisation process produced by the use of collaborative space, we elaborate research on socialisation in the context of inter-organisational construction projects, focusing on co-locational spaces and using illustrative cases of four alliance projects implemented in a Northern European country during 2011–2018. All the case projects used collaborative spaces as a key strategy in their project integration efforts. We illustrate the emerging socialisation tensions related particularly to (i) the spatial design of the collaborative space (ii) the facilitation of collaborative work in the collaborative space, (iii) the emergence of boundaries between full- and part-time members of the collaborative space, and (iv) the development of a shared identity in the collaborative space. The findings increase the understanding of the salient role of space in socialisation processes in inter-organisational construction projects.

11.2 Theoretical Background

11.2.1 The Origins of Co-Locational Spaces and Their Use in Construction Projects

Co-locating a team in the same physical space has become a popular workplace practice in the global landscape that has been identified as a salient factor in managing integration and coordination in different operational contexts and industries (Khanzode and Senescu 2012). In general, co-location refers to the use of a shared space where members, for example, from different units or organisations, are physically co-located and can interact in real-time face to face (Bosch-Sijtsema and Tjell 2017). In practice, co-location can take different forms, ranging from full-time co-location to recurring or staged co-locational activities. The concept of collaborative space is considered to consist of the physical space and socialisation and collaborative practices that take place in that space (Kokkonen and Vaagaasar 2018).

The origins of co-location can be traced back to the rise of lean thinking at Toyota in the 1990s. An 'Obeya' was set up to co-locate team leaders from different engineering areas to discuss problems, innovate, and make joint decisions with the objective of cutting lead times in automobile production and product development (Morgan and Liker 2006). Since then, the value of co-location in increasing knowledge sharing has been particularly emphasised in the new product development context (Coradi et al. 2015). In the space industry, the Jet Propulsion Laboratory of the National Aeronautics and Space Administration (NASA) is famous for co-locating an experienced design team, Team X, to complete extremely rapid designs with interdependent tasks routinely (Mark 2002). Lean thinking ideals have also affected the use of co-locational space and the emphasis on face-to-face interaction in agile projects, which have become the dominant form of delivering working software in the information technology (IT) industry (Hobbs and Petit 2017). In this context, fast communication and interaction between team members from different organisations are essential for quick responses to changes in the software.

The use of co-locational space has also spread to managing construction projects, particularly due to the use of collaborative procurement forms such as integrated project deliveries and alliance projects (Mosey 2019; Walker and Lloyd-Walker 2015). The rise of physical co-location in construction projects has also produced various concepts, including interactive workspace (Johanson et al. 2002), Big Room (integrated Big Room, intensive Big Room, virtual Big Room; Khanzode et al. 2008), and collaborative space (Kokkonen and Vaagaasar 2018). Each emphasises the role of face-to-face communication, the use of virtual technologies, the use of collaborative practices, and the frequency, intensity, and temporal orientation of physical co-location.

11.2.2 The Use and Outcomes of Collaborative Spaces in Inter-Organisational Construction Projects

Research on the role of collaborative spaces in construction projects is dominated by studies focused on identifying the benefits that the co-location of inter-organisational project team members may produce. Researchers have advocated that the activities carried out in a mutual work environment enhance collaboration between the project parties through increased interaction and communication (Khanzode and Senescu 2012). Some scholars have also presented co-locational teamwork as positively impacting negotiation behaviour (Raisbeck et al. 2010). Similarly, the development of mutual trust has been discovered to influence people's willingness to share knowledge and collaborate and, thus, have positive effects on project outcomes (Baiden et al. 2006). Working as a co-located team has been reported to promote team integration in multi-party construction projects, particularly between designers and constructors (Ibrahim et al. 2013). Furthermore, a team's co-location is presented as supporting design integration, such as the use of building information modelling and integrated concurrent engineering methods (Khanzode et al. 2008). Working collaboratively in a Big Room environment has been found to lower barriers to working with the latest information and more concurrently, which, in turn, speeds the project's progress (Khanzode and Senescu 2012).

Different kinds of collaborative project space set-ups have also been found to produce different types and degrees of collaborative behaviours and interaction outcomes: while interactive workspaces and intensive Big Rooms foster occasional but intensive cooperation through interactions with the virtual content and face-to-face collaboration of project members (Leicht et al. 2009), collaborative spaces and permanent Big Rooms can enable more in-depth formal and informal collaboration as the joint space facilitates continuous relationship building and collaborative working between the different disciplines (Kokkonen and Vaagaasar 2018; Lehtinen and Aaltonen 2020). Seeing other project members every day, organising regular meetings and working closely together as a result of co-location has been found to support the development of mutual trust in inter-organisational projects (Bygballe and Swärd 2019). Scholars have also examined how the use of facilitators and the implementation of behavioural norms in collaborative spaces can facilitate the formation of a joint collaborative identity in construction projects (Hietajärvi and Aaltonen 2017) and support the institutionalisation of collaborative routines within inter-organisational arrangements (Bresnen 2007; Bygballe and Swärd 2019). Particularly, the use of informal events and meetings and development of joint values and norms for the behaviour in collaborative spaces during the early project stages has been found to promote socialisation and trust development in the project organisation (Aaltonen and Turkulainen 2018).

Recently, the impacts of co-locational work on behaviour have been problematised, and the role of managerial practices and facilitators as important mediators for the realisation of benefits of collaborative spaces has been acknowledged (Kokkonen and Vaagaasar 2018). Some studies have also focused on the role of the spatial design of the workspace layout and how it may prevent social interaction in multi-party projects (Bosch-Sijtsema and Tjell 2017; Karrbom Gustavsson 2015).

11.2.3 Socialisation in Inter-Organisational Projects

Socialisation can be defined as 'the level of interaction between, and communication of, various actors within and between organizations, which leads to the building of personal familiarity, improved communication and problem solving' (Cousins and Menguc 2006, p. 607). It facilitates the building of inter-personal relationships, trust, interaction, and knowledge sharing, ultimately facilitating better performance (Cousins et al. 2008). Research on inter-organisational projects describes various collaborative practices for facilitating socialisation among the participating organisations, including workshops, relationship programmes, co-locational collaborative spaces, use of facilitators, and joint training (Bresnen 2007). However, the use of these practices has rarely been problematised.

The majority of the existing studies have focused on formal mechanisms for facilitating collaboration. However, there is growing interest in the social dimensions of collaboration and in understanding the role of informal socialisation practices in facilitating the emergence of trust and a collaborative mind set in projects 'naturally' (Bresnen 2007; Hietajärvi and Aaltonen 2017; Suprapto et al. 2015). Furthermore, we have limited understanding of how the spatial setting and the use of co-locational space shapes the socialisation process in inter-organisational projects.

11.3 Methodology

Contrary to the majority of previous research that has primarily addressed the role of co-locational spaces in advancing collaboration and cooperation, the aim of the study is to extend the understanding of the socialisation tensions that collaborative spaces may produce in an inter-organisational construction project. We adopt the theory elaboration research approach (Ketokivi and Choi 2014) and build on research on socialisation and existing understanding of co-location in the context of inter-organisational construction projects. We collected empirical data from four case projects and use the cases in an illustrative manner to sharpen and advance the understanding of socialisation tensions in the inter-organisational project context.

We collected qualitative data with an in-depth multiple case study method (Yin 2009). A case study approach was considered appropriate because it is particularly suitable for theory elaboration purposes (Ketokivi and Choi 2014) and assists in producing a rich analysis and understanding of complex phenomena.

As we were interested in understanding the project practices and activities and the socialisation processes taking place in collaborative spaces, we sought contemporary case projects that use collaborative space as their integration strategy. To control the potential variation that the project delivery form may bring to socialisation processes, we decided to focus on construction project alliances (Walker and Lloyd-Walker 2015). A project alliance can be considered an extreme form of relational integration in projects, where relational and behavioural aspects and collaborative work are emphasised to optimise the value creation of the project system. The selected replicative four case projects (Case Railway, Case Tunnel, Case Tramway, and Case Hospital) were all complex construction projects from an organisational and technical perspective, requiring emphasis on socialisation and collaborative work. We decided to focus on the use of collaborative space and the development of socialisation during the development phase of the project alliance. In this phase, the project team is formed, and the project is developed jointly by the alliance participants such as the client, designers, and main contractors. The use of the collaborative space and the socialisation dynamics are typically most intensive during this phase of the project.

Case Railway is a rail renovation project that includes three alliance organisations (development phase in 2011), Case Tunnel is a complex tunnel construction project in a city centre that includes five alliance organisations (development phase in 2015), Case Tramway is a light rail construction project that includes three alliance organisations (development phase in 2016), and Case Hospital (development phase in 2017) is a new and ambitious women's and children's hospital project that includes four alliance organisations, all executed in a Northern European country. The value of each project was around or more than 100 million euros.

Data collection regarding the use of collaborative space and socialisation processes focused on understanding how the project used Big Rooms in practice, what socialisation practices were in place, and how people experienced them. Understanding the potential tensions and conflicts that the alliance project participants had experienced in their projects in use of the Big Room and interaction within it was also important. The data collected (between 2014 and 2018) on all cases included 37 semi-structured interviews (7 at Case Railway, 11 at Case Tunnel, 15 at Case Tramway, and 4 at Case Hospital), non-participant observations in all Big Rooms, and research-related workshops. All interviews were recorded and transcribed as text. The data were complemented with project documentation, including organisational charts of the project, process descriptions, behavioural guidelines, and project reports. The number and format of the workshops varied between the cases; for example, in Case Tramway, extensive workshops were held, including one on the use of the Big Room.

The interviewees represented the key alliance organisations in all cases and were from different levels of the organisations, that is, from the leadership team, project management team, and project team. Individuals who worked full and part time in the Big Rooms were interviewed.

We began the data analysis by building a thorough understanding of the case projects, on the use of and behaviours and practices in the collaborative spaces, as well as on the tensions the project members experienced regarding use of the Big Room and socialisation in general. Then, we coded the interview transcripts, documentary data, observation memos, and workshop memos with Excel. During the data analysis, we observed that similar types of challenges and socialisation tensions were raised in all the cases. We realised that the use of the collaborative space was associated and posed challenges for the socialisation process in the temporary inter-organisational setting and that the project organisations were struggling with socialisation tensions that did not follow, for example, the traditional inter-organisational boundaries. The data also indicated that the project organisations struggled with finding the most suitable ways to use the collaborative space to facilitate socialisation. We found many indicators and instances in all the cases where the use of the collaborative space was linked to the outcomes of socialisation. We categorised the indicators into four themes that we labelled socialisation tensions produced during the use of the collaborative space.

11.4 Findings

We identified four themes that were linked to tensions in the use of the collaborative space and thus, also contributed to the development of socialisation in the inter-organisational construction projects.

11.4.1 The Spatial Design of the Physical Collaborative Space

Planning and designing the actual physical space that would serve as a collaborative space was not experienced as a trivial task by the project representatives. It was evident that the role the actual physical layout would play in the team's socialisation process was acknowledged by the managers. However, at the same time, they experienced that they did not have the proper tools or unified ideas on how to plan the optimal layout and the physical space. It seemed that the managers longed for simple best practice guidelines related to spatial planning but faced uncertainty and differing and even controversial ideals about the optimal solution. This was illustrated by a Case Hospital representative:

> We have still a lot to do with regard to the planning of the layout, and we need to modify it throughout the project and experiment and adjust. We are not completely sure what kind of layout works best in this project.

Consequently, the data indicated that although the interviewees could easily list the most important features of collaborative spaces, such as visual attractiveness, flexibility, and somehow distinctive from what they were used to, they were unsure about the connections in the physical space that fostered informal and formal collaborative behaviours among the different disciplines. In some cases, the interviewees seemed to assume that collaboration would emerge in the co-locational space without any formal planning of the layout, which caused socialisation-related tensions in the project. This was described by a Case Tramway coordinator:

> We have not paid enough attention to the planning of the spatial space in the beginning as it seemed that we just came here, eager to work with the project, and started sitting somewhere. The planning was not that systematic, and the assumption was that we just start collaborating.

In all case projects, the project managers and coordinators emphasised cross-disciplinary collaboration between designers and builders in the early development phase, but at the same time, they acknowledged the need to advance design processes efficiently among the designers. This led to dilemmas in space layout planning, as the projects struggled with whether to mix professionals from different disciplines or, for example, locate structural designers in one place and facilitate collaborative work through other means such as workshops. At the same time, the interviewees were afraid that cliques would form because of a non-optimal space layout where, for example, people from the same company or design discipline sat next to each other. These tensions introduced layout planning dynamics: 'We have experimented pair-working among the designers and builders and also made some adjustments to the space layout use in this open space to facilitate cross-disciplinary working' (Case Tunnel). The space layout and use, therefore, were dynamically adjusted throughout the development phase to deal with the tensions and emerging project needs.

The ideal of open-space layout was visible in all the cases. Each project had some kind of large open-space office with designated workstations. The meeting and workshop facilities were either in the middle of the open space or in separate rooms. Some project members strongly believed that information and knowledge would transfer fluently in the open space as people could easily hear, for example, what the managers were discussing in the meetings, and one could tap a colleague's shoulder and ask for more information for some emerging problem. However, although it was evident in all the case projects that the open-space planning facilitated information sharing, the project members also struggled considerably with this configuration. Many interviewees expressed concerns and problems with 'being available all the time' and the noise levels. Interviewees also felt that it was not always motivating to listen to the conflicts and challenges discussed in the project team meetings. Some individuals even mentioned that if they wanted to really concentrate on their work, they did not necessarily come to the collaborative space, but went to their parent organisations' spaces to work. Therefore, the multi-tasking approach that the open-space layout produced was also a factor that decreased work efficiency. Furthermore, individuals questioned the assumptions that knowledge transfer could be facilitated simply through open-space layout design. Interviewees felt that in some cases the belief in 'knowledge transferring by itself' through co-location was so strong that not enough attention was paid to facilitating knowledge transfer through managerial activities or knowledge IT systems. This was illustrated by a Case Railway representative:

> I need to have a reason for coming to the co-location space. Otherwise, it is not productive for me to spend time there, as work can be distracted. I see some value in socialisation, but at the end of the day, it is still about doing those things that advance the project, not socialising.

Tensions and struggles due to the spatial design and the socialisation process were also experienced by individuals, particularly designers, who were working on multiple projects at the same time. Typically, agreements were made that individuals could work on other projects while co-located in one project's co-locational space. This was a managerial strategy to make sure that certain individuals were available face to face when needed. However, some of the design professionals felt that this meant they should be available all the time. They also mentioned that the spatial layout did not address working on multiple projects at the same time. It was not very easy to work on other projects, particularly in the open space, if there was information that should be kept secret, and if meeting rooms were unavailable. The data revealed that the project members experienced that there were not enough meeting rooms for individuals to work, and their use rate was high. To signal and symbolise transparency, in many cases, the doors were always kept open, even when meetings were held. However, this led to closed doors raising suspicions.

Tensions also emerged regarding the use of walls and visualisation in the spaces. The emphasis on visualisation increased in all the case projects during the development phase, but its usefulness was valued differently by the interviewees. Some emphasised the importance of sharing information through visual elements such as innovation spreadsheets or valued visual organisational charts. However, in other projects, the wall space was dedicated to only the Last Planner system. Its physical versus virtual presence was also constantly debated by the project members. Some individuals felt that its use should only be physical, but some perceived that kind of use as very outdated and did not place much value on the social processes related to its use.

Finally, the planning process of the spatial design produced some tensions among the project members. Interviewees pointed out that a separate Big Room coordinator position should be established, but others favoured a continuously self-adapting planning process. The questions of who could be and should be in charge of planning the collaborative space in an inter-organisational project and how widely project members should be engaged in this process were indicators of power struggles related to the collaborative space.

11.4.2 Facilitation of Collaborative Working in the Collaborative Space

The actual collaborative working and collaborative practices within the collaborative space were constantly debated and entailed tensions that were not all resolved during the projects' development phases. The most significant differences in perceptions were related to the formalised versus informal approach to collaboration in the co-locational space. Some interviewees were very much in favour of 'engineering collaboration' within the space and favoured systematic planning and scheduling of the space use and workshops. Others felt that this approach would lead to too inflexible structures that would not produce value for the collaborative work in the space. The need to balance these two is described by a Case Hospital representative:

> We need to plan very systematically the weekly timetables and the activities what to do in the Big Room when different individuals come and go. Then, on the other hand, in order to allow ideation and encourage creativity, we should be able to offer more relaxed and informal moments. This is some kind of a challenging balancing process.

This debate was also intertwined with the use of dedicated and specialised collaboration facilitators: Some managers and project members called for a separate collaboration coordinator, who would be assigned tasks related to collaboration planning and execution. Others believed that relying too much on the formalised rules of collaboration and separate positions could destroy the project team's innovativeness and ideation capability. 'I do not know if you can really outsource the task of collaborating, it is not a task of someone. It is a joint process, and formalising it does not really pay off', stated a Case Tramway representative. In addition, the debate over whether internal facilitators would be better than external consultants was also highlighted, as well as the struggles regarding which project party would be in the right position to act as a facilitator.

The evidence also points to a constant struggle of inclusiveness versus exclusiveness in the use of the collaborative space that had implications for the socialisation process. Typically, at the start of the development phase, the managers highlighted an inclusive approach to engage the project personnel as widely as possible in the different project procedures and to facilitate commitment. This approach led to meetings and, for example, Last Planner sessions that were long and extensive and produced experiences in which individuals felt that their presence in the meetings in the collaborative space was unnecessary. The exclusiveness approach was described by a Case Tunnel representative:

> At the start of the project, we were so eager to engage and start working broadly. Almost everybody was invited into the meetings in our co-locational space to ideate, which did of course result in inefficient meeting practices. Then we adjusted, for example, the number of people participating in Last Planner sessions so that we would have only those people who would contribute.

We observed a pattern in the data that after a period of inclusiveness during the early development phase, adjustments were made in terms of collaborative practices, typically resulting in exclusiveness strategies for the use of the collaborative space. Then, the strategies produced experiences of isolation among individuals who felt that they could not participate fully in the activities within the collaborative space.

11.4.3 Emergent Boundaries Between Full- and Part-Time Members of the Collaborative Space

The data also revealed an interesting emerging pattern in the organisational boundaries and the use of the collaborative space. As cooperation among the members of the project alliances developed in the collaborative space, the boundaries between the organisations started blurring. This was facilitated by a socialisation process of individuals who worked full time in the Big Room. At the same time, a novel boundary emerged based on whether individuals were located full or part time in the collaborative space. In other words, we noted a gradually emerging strong boundary between the full- and part-time members of the collaborative space. A unique 'collaborative space community' among full-timers who worked at the distinctive home base space of the project formed, which also reinforced the feelings of isolation among those who visited or used the collaborative space only occasionally or part time. This was illustrated by a Case Tramway representative as follows:

> I also feel that it is about the attitude. If I imply that this co-location space does not work people say that you should be here more. But not everybody can be here. They are hiding behind it that everybody should be here. It does not go like that. It is almost like, 'if you are not here three days a week, you are a bad person'.

This project core and periphery structure and us-versus-them mentality between those who were the true space users and those who were outsiders to a certain extent was an emerging and unintended consequence of the co-locational space strategy for the managers. The data showed that it can to some extent be explained by the unique space set-ups, and that the individuals seemed to identify with and the project's existence very strongly with the existence of and time spent in the physical space. This was also reinforced by the strategies that highlighted the Big Room as a key strategy for transferring knowledge. In other words, if you were not constantly present in the collaborative space, you were not necessarily considered a committed and true team member.

The part-time members, particularly designers, also experienced a very clear division between the users of the space. Some felt guilty if they were not physically within the space and felt that they were somehow blamed if they did not come to work in the space but worked in other co-locational spaces or in their company's office. In practice, they also felt challenges in terms of knowledge sharing: As the idea of the collaborative space was related to the free flow of information across those who were in the space, many times, the full-timers did not make an extra effort to share information with non-space users. Instead, the full-timers stated that it was the part-timers' own fault if they did not come to the space. These feelings were described by Case Tramway and Case Tunnel representatives:

> It is truly a problem that there are these kinds of unexpressed expectations that you should be so much physically co-located in Big Room to receive information and knowledge. Why do we not just admit that this is simply not possible for all of us and quit this game of playing that this would somehow be possible. If we do not admit it, development is not possible.
>
> There are these expectations that you should be in the Big Room, and you feel guilty if you cannot be physically present there. It is the place where information is transferred, and you can ask for input from people. It almost makes you feel that communication through other channels such as e-mail or phone in this project would not somehow be right. This can cause problems in information exchange.

Thus, the data revealed insights into how co-locational spaces and presence within them can be used as means of control and power in inter-organisational projects. They seem to enable new kinds of power structures in projects that are not tied to inter-organisational designs. Some informants felt that they were in a way 'punished' and missed some critical knowledge because they were not present in the co-locational space, as illustrated by a Case Railway representative:

> I know that I must attend the weekly meetings as planned and be there physically; otherwise, I will miss some information related to the project and, for example, its schedule, which is there on the wall in the form of a Last Planner.

Interviewees also brought up that calling or sending an e-mail to people who are part-time workers becomes increasingly challenging in cases where you have the collaborative space in place because it seemed that it was easy to decide to postpone contacting to the moment when the person is in the place physically. This indicates that on some occasions, the very essence of collaborative space can also decrease information exchange and socialisation in an inter-organisational project.

11.4.4 Development of a Shared Identity in the Collaborative Space

The fourth theme of socialisation tensions was related to the development and dynamics of the shared collaborative space identity and the means to facilitate it. The collaborative space identity was related to discussions about what the project space represents, how it differentiates the project from other projects, and how people should collaborate and behave in this project's co-locational space when compared to other projects. Struggles were also related to the activities through which the Big Room's spirit could be transferred to the sub-Big Rooms on the construction sites. These tensions were also tied to the emergence of boundaries between full-timers and part-timers as well as to struggles in terms of the collaborative processes.

The data indicate that some of the interviewees felt strongly that the project organisation should try to establish a unified and permanent collaborative space identity, for example, through codified rules for collaborative space and cooperation within it. Within this approach, engaging individuals at the start of the project to develop collaborative rules were favoured, as illustrated by a Case Tramway representative:

> It is important to build a joint culture and alliance spirit for the project and to try to establish the rules and values of this Big Room so that people know how they should behave and co-operate here. This also supports the socialisation of the newcomers to the Big Room.

On the other hand, some interviewees – a minority, however – considered this approach unrealistic and promoted a more fluid and polycentric approach to the culture and identification. They considered that the culture is always in constant flux, and that project members come and go. Therefore, attempts to invest too much in a shared culture do not pay off. Interviewees also experienced so many rules and demands for collaboration that are partly contradictory and produce complexity that knowing which to follow is often challenging. For example, in Case Tramway, there were rules and requirements for behaviour in Big Room, virtual etiquette, the client's requirements, and service promises that all entailed at least some kind of guidelines on how to collaborate in the co-locational space. Furthermore, socialisation tensions related to the collaborative unified identity were experienced most strongly by part-time workers, who often had other project commitments. As part-timers juggled multiple project-related identities, they found it challenging to attach themselves strongly to any project's identity. In addition, tensions increased regarding attempts to copy the central co-locational space's alliance spirit and identity in sub-Big Rooms on construction sites.

We also identified tensions related to project transformation phases. The strongest tensions were grievances related to the potentially changing 'team spirit' within the collaborative space. These instances were evident when the physical collaborative space was changed as the project progressed and changes in the team were made. Interestingly, the evidence indicates very strong mental attachments to the actual physical space and its community that symbolised and represented the project for many people, as illustrated in the following comments: 'The Big Room is more than a physical place. It is a mental state of our project' (Case Tunnel). 'It has been such a great journey that you are already a bit afraid that it will end' (Case Railway). When the projects entered the execution phase, the collaborative spaces were almost always dispersed to the construction sites, and the project team needed to recreate its identity. A lot of work to facilitate creating shared identity was related to this phase.

11.5 Discussion and Conclusions

The results broaden and problematise the existing understanding of the role of co-location spaces in producing favourable outcomes in inter-organisational construction projects as well as provide ideas for managers on how to use collaborative spaces in their projects to facilitate socialisation.

11.5.1 Theoretical Contribution and Implications

The key contribution of the study is the identification of four novel tensions that the use of collaborative space may produce in the socialisation process of inter-organisational construction projects. The identified tensions shed light on the potential 'dark sides' and struggles that inter-organisational projects may face through the use of collaborative spaces. Overall, the results portray a more fine-grained picture of socialisation processes in inter-organisational projects that extends beyond the core team (Aaltonen and Turkulainen 2018) and provide evidence of the potential complexities associated with the use of collaborative space and realisation of their benefits (Bosch-Sijtsema and Tjell 2017; Kokkonen and Vaagaasar 2018). Consequently, the findings of the study suggest that the path towards favourable collaboration may advance through tensions and struggles that managers must resolve to realise the value of collaborative spaces.

The findings related to the design of spatial space reveal challenges in planning the layout (Bosch-Sijtsema and Tjell 2017). Similarly, the results related to the facilitation of teamwork lend support to previous findings of a need for facilitation and coordinators in collaborative spaces (Karrbom Gustavsson 2015). To some extent, the results also question the appropriateness of formalised engineering of collaboration in all phases of the project, resonating with the findings for collaboration paradoxes (Bresnen 2007). The present empirical evidence revealed how organising collaboration in a collaborative space may require time from the project team and that controversies exist regarding finding the right balance between formalisation and informal collaboration across the whole project team, and how this may foster socialisation. Socialisation, therefore, has a temporality to it.

The findings related to the emerging boundaries between the full-time inhabitants of the collaborative space and others on the periphery provide a novel perspective that incorporates the role of the physical space when analysing organisational boundaries in inter-organisational projects. To date, much research has focused on inter-organisational boundaries and how they affect cooperation and coordination in projects (Oliveira and Lumineau 2017). The present results suggest that this might not always be the case, particularly in project alliance contexts, where the physical space seems to play a significant role in the formation of boundaries and in project identification processes. The evidence also suggests that the physical space can be utilised as a powerful means of exercising power, which is a relatively novel perspective on inter-organisational behaviours. Paradoxically, the best intentions to facilitate integration and knowledge transfer through the use of the collaborative space may result in frustrations in some parts of the project; at worst, the notion of openness may then produce isolation.

Finally, the findings for identity tensions and struggles advance the notion and understanding of the role of collaborative space in promoting the formation of inter-organisational project identity (Hietajärvi and Aaltonen 2017) by showing how the use of collaborative space may pose identity struggles and affect socialisation processes. The notion of space as a producer of identity-related tensions sheds light on the potential dark side of collaboration, as researchers have focused primarily on the positive mechanisms through which a co-locational space can advance a joint alliance spirit (Walker and Lloyd-Walker 2015).

The findings on the countervailing requirements of the members and socialisation tensions also reveal important features of project organising and organisational design in general. First of all, organising in a collaborative space is an ongoing and non-linear process where balancing and mastering of contradictory or even paradoxical organising requirements. This view challenges and is in sharp contrast with the more static and linear approaches concerning the organisational design of projects as suggested by Schreyögg and Sydow (2010). Socialisation processes and organising in general are also unfolding throughout the project lifecycle: the degree of socialisation is constantly developing in the project organisation and the organising solutions need to adapt to this change. Balancing between self-organising processes allowing flexibility and more formal team management processes including, for example, planned collaboration facilitation is, therefore, essential. For example, self-organising processes were favoured by some of the team members during the early stages of the projects to promote flexibility, spontaneous interaction and improvised processes, while formal team management approaches and the development of norms and values and routines through planned and facilitated collaborative workshops were required by other members to systematise collaborative work later on. The co-location of the project team also makes dysfunctional organising processes and challenges more visible to the inter-organisational project team members and may enable rapid adjustments of organising solutions, which do not seem to work as well as more flexible division of responsibilities among the project team members. Different types of integrative cross-disciplinary working groups were, for example, added in one of the projects, when the collaborative working between the disciplines did not work. As the findings demonstrate, the use of collaborative spaces also blurs organisational boundaries and requires a shift from the firm centred project management approaches to the management of project teams with representatives from different organisations (Hietajärvi and Aaltonen 2017). Consequently, instead of strengthening and developing the collaborative identity of each organisation participating in the project, focus in organising is shifting to the development of a joint and common, but temporary, collaborative identity within the single project.

11.5.2 Further Research

This elaborative study used four alliance cases as a means of illustrating socialisation tensions produced by the use of collaborative spaces. The study focused on cases that used collaborative spaces for integrating project personnel during the development phase of project alliances. Studies on the use of part-time co-locational spaces and on different project life-cycle phases might produce different outcomes. Further research should also pay attention to the managerial strategies, responses, and balancing acts through which the illustrated socialisation tensions are resolved in different inter-organisational projects. This would increase the understanding of how the responses are contingent on the emerging socialisation tensions and, thus, facilitate the development of contingency-based understanding of inter-organisational construction projects and their management. Research could also focus on developing models for the assessment of positive and negative outcomes (net benefits) of co-location in different kinds of projects. Understanding better how organisational behaviour and relationship management in particular is implemented in co-locational spaces would provide a valuable knowledge of collaborative dynamics in inter-organisational projects. Finally, the study opens up new avenues for further research related to the links between co-locational space and identity formation in construction projects, as well as to the emerging boundaries between the core and periphery determined by the user status of the co-locational space.

11.5.3 Managerial Implications

The findings for the use of collaborative space and their links to the socialisation processes in inter-organisational construction projects offer managers various key lessons. First, managers should be aware of the potential tensions that the use of co-locational space may produce in project teams, be able to reflect upon the tensions, and formulate strategies for managing such conflicts in their projects. The path to realisation of the benefits of co-locational space is complicated and requires extra investment in managing collaborative behaviours. Managers should plan the layout of collaborative spaces carefully by addressing the challenges experienced by part-time workers and be able to adjust the layouts when needed. Transition phases from one project stage to another are particularly important for the redesigning of the layout and collaborative practices. Layout planning should consider the need for designer–contractor interaction as well as ensure that members from different organisations are mixed. Managers should also be tuned to observing signals about the potential formation of cliques of full- and part-time workers in co-locational spaces. Regular workshops that encourage project members to discuss and reflect upon the co-locational space working-related challenges and come up with improvement solutions are needed. Efforts should be directed to more carefully integrate the periphery employees into the project, and collaborative space should not be used as the only means of knowledge transfer. To manage this challenge, virtual solutions to integrate part-time members and ensure their presentism are recommended. Facilitating collaboration and collaborative practices should also be addressed with care, and in some projects, a specific role of collaboration facilitator can be useful. Enough time should also be reserved for developing collaborative practices in the co-locational space together, which facilitates the development of joint norms, values, and identity. Furthermore, identity struggles among part-timers should be considered in the strategies for project identity formation.

References

Aaltonen, K. and Turkulainen, V. (2018). Creating relational capital through socialization in project alliances. *International Journal of Operations & Production Management* 38 (6): 1387–1421.

Baiden, B.K., Price, A.D.F., and Dainty, A.R.J. (2006). The extent of team integration within construction projects. *International Journal of Project Management* 24 (1): 13–23.

Bosch-Sijtsema, P.M. and Tjell, J. (2017). The concept of project space: studying construction project teams from a spatial perspective. *International Journal of Project Management* 35 (7): 1312–1321.

Bresnen, M. (2007). Deconstructing partnering in project-based organisation: seven pillars, seven paradoxes and seven deadly sins. *International Journal of Project Management* 24 (4): 365–374.

Bygballe, L.E. and Swärd, A. (2019). Collaborative project delivery models and the role of routines in institutionalizing partnering. *Project Management Journal* 50 (2): 161–176.

Coradi, A., Heinzen, M., and Boutellier, R. (2015). Designing workspaces for cross-functional knowledge-sharing in R&D: the 'co-location pilot' of Novartis. *Journal of Knowledge Management* 19 (2): 236–256.

Cousins, P.D. and Menguc, B. (2006). The implications of socialization and integration in supply chain management. *Journal of Operations Management* 24 (5): 604–620.

Cousins, P.D., Lawson, B., and Squire, B. (2008). Performance measurement in strategic buyer-supplier relationships: the mediating role of socialization mechanisms. *International Journal of Operations & Production Management* 28 (3): 238–258.

DeFillippi, R. and Sydow, J. (2016). Project networks: governance choices and paradoxical tensions. *Project Management Journal* 47 (5): 6–17.

Gupta, A.K. and Govindarajan, V. (2000). Knowledge flows within multinational corporations. *Strategic Management Journal* 21 (4): 473–496.

Hietajärvi, A.M. and Aaltonen, K. (2017). The formation of a collaborative project identity in an infrastructure project alliance. *Construction Management and Economics* 36 (3): 1–21.

Hobbs, B. and Petit, Y. (2017). Agile methods on large projects in large organizations. *Project Management Journal* 48 (3): 3–19.

Ibrahim, C.K.I.C., Costello, S.B., and Wilkinson, S. (2013). Development of a conceptual team integration performance index for alliance projects. *Construction Management and Economics* 31 (11): 1128–1143.

Johanson, B., Fox, A., and Winograd, T. (2002). The interactive workspaces project: experiences with ubiquitous computing rooms. *IEEE Pervasive Computing* 1 (2): 67–74.

Karrbom Gustavsson, T. (2015). New boundary spanners: emerging management roles in collaborative construction projects. *Procedia Economics and Finance* 21: 146–153.

Ketokivi, M. and Choi, T. (2014). The renaissance of case research as a scientific method. *Journal of Operations Management* 32 (5): 232–240.

Khanzode, A. and Senescu, R. (2012). *Making the Integrated Big Room Better. DPR Construction*

Khanzode, A., Fischer, M., and Reed, D. (2008). Benefits and lessons learned of implementing building virtual design and construction (VDC) technologies for coordination of mechanical, electrical, and plumbing (MEP) systems on a large healthcare project. *Journal of Information Technology in Construction (ITcon)* 13 (22): 324–342.

Kokkonen, A. and Vaagaasar, A.-L. (2018). Managing collaborative space in multi-partner projects. *Construction Management and Economics* 36 (2): 83–95.

Lehtinen, J. and Aaltonen, K. (2020). Organizing external stakeholder engagement in inter-organizational projects: opening the black box. *International Journal of Project Management* 38 (2): 85–90.

Leicht, R.M., Messner, J.I., and Anumba, C.J. (2009). A framework for using interactive workspaces for effective collaboration. *Journal of Information Technology in Construction (ITcon)* 14 (15): 180–203.

Mark, G. (2002). Extreme collaboration. *Communications of the ACM* 45 (6): 89–93.

van Marrewijk, A., Ybema, S., Smits, K. et al. (2016). Clash of the titans: temporal organizing and collaborative dynamics in the Panama Canal megaproject. *Organization Studies* 37 (12): 1745–1769.

Morgan, J.M. and Liker, J.K. (2006). *The Toyota Product Development System*. New York: Productivity Press.

Mosey, D. (2019). *Collaborative Construction Procurement and Improved Value*. Hoboken, NJ, USA: Wiley Blackwell.

Oliveira, N. and Lumineau, F. (2017). How coordination trajectories influence the performance of interorganizational project networks. *Organization Science* 28 (6): 1029–1060.

Puranam, P. (2017). When will we stop studying innovations in organizing, and start creating them? *Innovations* 19 (1): 5–10.

Puranam, P., Alexy, O., and Reitzig, M. (2014). What's 'new' about new forms of organizing? *Academy of Management Review* 39 (2): 162–180.

Raisbeck, P., Duffield, C., and Xu, M. (2010). Comparative performance of PPPs and traditional procurement in Australia. *Construction Management and Economics* 28 (4): 345–359.

Schreyögg, G. and Sydow, J. (2010). Organizing for fluidity? Dilemmas of new organizational forms. *Organization Science* 21 (6): 1251–1262.

Suprapto, M., Bakker, H.L.M., and Mooi, H.G. (2015). Relational factors in owner-contractor collaboration: the mediating role of team working. *International Journal of Project Management* 33 (6): 1347–1363.

Van Maanen, J. and Schein, E.H. (1979). Toward a theory of organizational socialization. *Research in Organizational Behavior* 1: 209–264.

Walker, D.H.T. and Lloyd-Walker, B.M. (2015). *Collaborative Project Procurement Arrangements*. Newtown Square: Project Management Institute.

Weick, K.E. (1974). Amendments to organisational theorizing. *Academy of Management Journal* 17 (3): 487–502.

Yin, R.K. (2009). *Case Study Research – Design and Methods* (Revised ed.). Thousand Oaks: Sage Publications.

12

On the Synchronisation of Activities During Construction Projects

Tuomas Ahola

Professor of Industrial Engineering and Management, Tampere University, Finland

12.1 Introduction

When the achievement of a shared goal depends on the collaboration of multiple individuals or organisations, the synchronisation of activities carried out by the participants is often of crucial concern. For example, the outcomes of work carried out by teams of medical professionals working in emergency rooms, members of jazz orchestras, special weapons and tactics officers, or Olympic athletes running a 400-m relay are highly dependent on the individual participants' ability to work together, seamlessly synchronising the interdependent activities carried out by members of the team.

Construction projects are inter-organisational arrangements established to complete a large network of tasks characterised by complex interdependencies. At a practical level, subcontracting is used to distribute tasks to heterogeneous firms assuming diverse roles, such as the main contractor, subcontractor, and supplier of components. To minimise the time wasted on waiting for other firms to complete their tasks, it is crucial that such interdependent activities remain synchronised throughout the course of the project. However, construction projects are characterised by a variety of problems, including poor activity–duration estimates, logistical problems, quality problems, and scarcity of competent human resources, that frequently result in missed deadlines (Chan and Kumaraswamy 1997; Assaf and Al-Hejji 2006). Due to task interdependencies, problems encountered in a specific task frequently hinder the progress of other firms completing other tasks. To counter this, the resourcing and scheduling of tasks often need to be continuously adjusted by the organisations involved in the construction project. This process of organising is complex and dynamic and spans across organisational boundaries. Consequently, the coordination of interdependent tasks is often suboptimal, unfavourably affecting project performance as well as quality of outputs. In some projects, delays cascade and spread uncontrollably across organisational boundaries.

This chapter focuses on the synchronisation of activities in inter-organisational construction projects. First, drawing on prior literature, I discuss the types of task interdependencies

Construction Project Organising, First Edition. Edited by Simon Addyman and Hedley Smyth.

exhibited by such projects and how they are generally managed. Next, relying on interviews I have carried out with project professionals, I elaborate on the processes by which the synchronisation of interdependent activities is lost and restored. I then discuss how the synchronisation of activities is associated with performance outcomes at the level of both construction projects as a whole and the individual organisational participants. Towards the end of this chapter, I examine factors that support the ability of individual organisations and the entire inter-organisational project network to maintain synchronisation over the course of the project and practices/strategies project managers use to restore synchronisation.

12.2 Synchronisation of Activities in Construction Projects

12.2.1 Task Interdependencies and Synchronisation

Determining how to achieve and maintain the synchronisation of interdependent tasks in production processes involving multiple actors represents a core challenge of organising such processes, whether they are intra- or inter-organisational in nature. Here, 'synchronisation' refers to the timing and coordination of interdependent tasks in a manner that reduces the amount of non-productive working time (for example, time wasted on waiting for another task to be completed or on the rescheduling of activities). In production operations, interdependencies between tasks may be pooled, sequential, or reciprocal in nature (Thompson 1967; Donaldson 2001). In pooled interdependence, two or more actors make a distinct contribution to a shared goal. A job shop is a frequently discussed example of a production process with pooled interdependence (Galbraith 1973). In this case, each worker can rather independently concentrate on their tasks, such as the assembly of products produced in the shop. No coordination between individual workers is typically necessary, but the outputs of all the workers contribute to the organisation's goals. In construction projects, supply chains for materials typically required in construction work (such as concrete and wood) exhibit pooled interdependencies, and exceptionally high demand or disruptions in supply can reduce the availability of critical materials.

In sequential interdependence, the output of one task serves as the input of another. In manufacturing environments, sequential interdependencies frequently characterise production lines and bottlenecks in batch processes (Woodward 1965). It has been argued that in the construction industry, the traditional production process is a classic case of the sequential interdependence of work (Winch 1989). The efficiency of projects involving a high number of sequentially interdependent tasks is supported by the detailed planning of schedules and resources and the order in which activities are to be carried out. This issue lies at the heart of project-management research. Reciprocal interdependence, in which the resources and actions of participants working on a task may affect the resources or actions of other participants, is both the most complex form of interdependence and the most difficult to coordinate (Galbraith 1973; Walker 2007). In construction projects, both the design and installation of various interdependent systems, such as electrical, heating, and ventilation, frequently require ongoing coordination and adjustment from several participating individuals or organisations. Although similar to sequential interdependencies, reciprocal interdependencies can also be partially coordinated by sophisticated planning techniques, the possible interactions are simply too numerous for such techniques to fully address the problem (Tushman and

Table 12.1 Management of task interdependencies in construction projects.

Type of interdependence	Pooled	Sequential	Reciprocal
Examples of construction projects	Material supply chains, work tasks of low complexity (e.g. painting)	Most tasks during the implementation phase	Many design activities, installation of subsystems in constructed building
Need for task synchronisation	Low	Moderate	High
Examples of managerial approaches for maintaining the synchronisation of activities	Synchronisation typically unnecessary (resource planning, purchasing)	Resourcing techniques and scheduling	Frequent communication, shared working practices and tools, co-location, joint problem-solving approaches (mutual adjustment)

Nadler 1978). In addition, Tushman and Nadler (1978) suggested mutual adjustment as a means for further alleviating the problem of reciprocal interdependence. In mutual adjustment, any involved actor may introduce new information related to the coordination of the task in question at any time. As such, mutual adjustment closely resonates with the dynamic and process-oriented ways of organising discussed in modern organisation theory. Finally, the challenges associated to reciprocal interdependence may also be tackled by the adoption of modularity in both organizational interfaces as well as product structures, as they have been argued to reduce the cost and difficulty of coordination (Sanchez and Mahoney 1996). Table 12.1 summarises the management of task interdependencies in construction projects.

12.2.2 Synchronisation of Activities in Temporary Organising

Although synchronisation is important to any productive task, its importance seems paramount in temporary organising, because time is by definition a scarce resource and generous safety margins cannot be incorporated into individual tasks to facilitate synchronisation (Lundin and Söderholm 1995). In addition, by contrast with process-based industries – such as petrochemical production and manufacturing-oriented industries with relatively stable supply chains as found in automotive production – the construction industry and other project-based industries are characterised by highly unique deliverables (Mandják and Veres 1998). Consequently, the chains of activities required to produce a construction project's deliverables are also highly unique. In addition, the inter-organisational networks of participating firms are established separately for each project (Hellgren and Stjernberg 1995), so the networks frequently include firms that have never worked together and, thus, have not developed effective ways to synchronise their production activities (Eccles 1981). The first meetings held at the start of the project have been argued to have a particularly central role for the coordination of activities, as behavioural pattens and assumptions guiding work emerge during them and become shared by the team members (Gersick 1988).

Construction projects are established for a specific purpose, like the construction of a shopping centre. Such projects involve the creation of a temporary project network, which ceases to exist immediately when the project is completed (or which is abandoned when projects fail). There are no guarantees that organisations participating in a specific construction project will be selected for subsequent projects, which means that the organisations must deal with a high degree of uncertainty regarding future opportunities, that is, a very limited shadow of the future. It can be assumed that under this condition, the actors involved in construction projects may be less concerned about the synchronisation of tasks across the project organisation than are organisational actors in a more stable network structure, such as that associated with automobile manufacturing (Dyer 1997) and apparel manufacturing (Uzzi 1997). However, previous research has shown that despite a limited view of the future, construction projects are often characterised by a high degree of structural embeddedness because many of the participating firms may have worked with each other in the past (Eccles 1981). Some of the participating actors may, thus, have developed shared practices and routines for working together effectively and maintaining or restoring the synchronisation of activities.

12.3 Processes by Which Synchronisation Is Lost and Restored

How is the synchronisation of interdependent activities in construction projects lost? A project manager discussed the progress of work in a technical area:

> We were supposed to carry out our work as we had done in the previous project together, but then it turned out that these subcontractors were so late, and they were missing all kinds of installations. And then when they started doing the installation work, and they managed to burn the paint so that it looked really bad. We had just started our work when the [client's] inspectors arrived and stopped our work. For a while, it was very unclear how we should proceed.
>
> (Project manager)

This example highlights that under the condition of task interdependence, the problems faced by one organisation can give rise to problems for another. As the project manager went on to explain, the desynchronisation process can be contagious:

> It was as a snowball would have been rolling down a hill. Any area I went to inspect, somebody was saying to me that 'we cannot finish this area as we are lacking the ordered furniture'!
>
> (Project manager)

> So, was the bottleneck the [main contractor's] factory?
>
> (The author)

> Absolutely. They had a lot of overlapping project deliveries at that time.
>
> (Project manager)

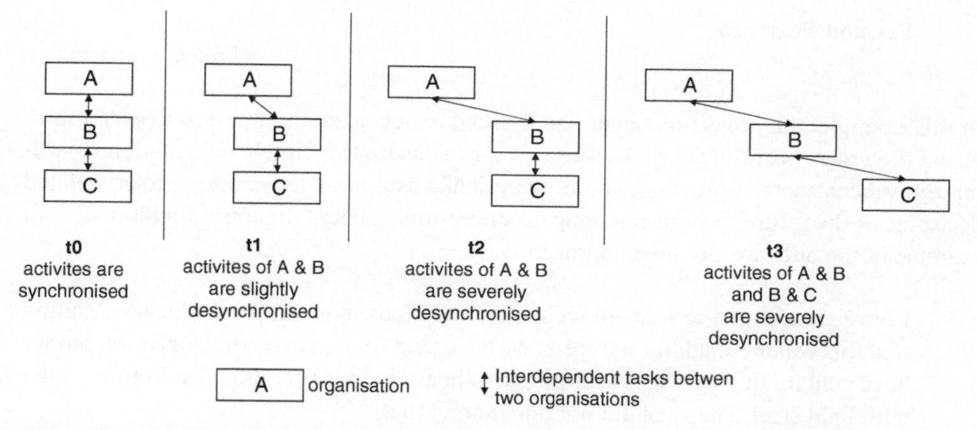

Figure 12.1 Process of desynchronisation in inter-organisational projects.

Figure 12.1 illustrates the process of losing the synchronisation of activities in an inter-organisational project. At t1, the interdependent tasks carried out by two organisations (A and B) are slightly desynchronised because A's work has fallen behind schedule, consequently slowing down B's work. At t2, the activities of the two organisations have become severely desynchronised, and B's work must be completely stopped until A completes its work. At t3, C's activities have become severely desynchronised with those of B.

The process of desynchronisation is often gradual, and it is difficult to identify a specific point in time when it begins; however, the point when work must be stopped is typically documented as stoppage and may have unfavourable consequences defined in the project contract for at least one of the involved parties. Severe desynchronisation does not always follow slight desynchronisation. Instead, either A or B, or both actors jointly, may deploy various measures to restore synchronisation, such as securing an additional workforce to speed up task completion. But how do organisations restore the synchronisation of tasks in inter-organisational projects? According to interviews carried out by the author, the measures adopted by organisations can be categorised as either inward-oriented or outward-oriented approaches. Inward-oriented actions aim to cushion the internal activities of the affected organisation from adverse conditions in its inter-organisational environment, whereas outward-oriented actions aim to influence the actions of organisations with which the affected organisation engages in interdependent tasks. A project manager discussed an inward-oriented approach:

> The pace of the project has increased during September and October, resulting in additional work for us. There are many instances during which we have to make sure that the [subcontractor] actually has 15 men working . . . as it claims. We have to take additional measures to verify that this is actually the case.
>
> (Project manager)
>
> So, would it be correct to ensure that the need to monitor the work of [the subcontractor] has increased in recent times?
>
> (The author)

> Yes, and clearly so.
>
> (Project manager)

In this example, the project manager had decided to devote additional resources to monitoring the progress of the delayed subcontractor to ensure that timely information regarding the subcontractor's progress would be available as a basis for making project-related decisions in the future. A project manager representing a client organisation discussed an example of the outward-oriented approach:

> I have asked them to redo all welds, but they have not complied. They are stating that this will incur additional costs. We have paid [the main contractor] once, but we have paid for quality work. Working with heavy steel is very different from working with light steel. They just did not understand that.

Furthermore, depending on the degree of desynchronisation (slight or severe) and the responses available to the actor, approaches to responding to desynchronisation events aim either to mitigate the event's harmful effects or to avoid the effects altogether. Figure 12.2 highlights the four types of approaches that may be available to project actors.

Firms may combine various inward- and outward-oriented approaches in their efforts to restore the synchronisation of tasks. A nascent stream of literature is devoted to uncovering the responses of actors when the synchronisation of tasks has been lost. Hällgren and Maaninen-Olsson (2005) showed that actors may search for information, change the composition of the project organisation, and engage in additional communication to restore the synchronisation of activities. More recently, Söderholm (2008) has demonstrated that project actors may apply detachment strategies and renegotiate the conditions under which tasks are carried out. Drawing on the Manhattan Project, Lenfle and Loch (2010) emphasised the importance of improvising and learning-by-doing in projects for which following

Orientation of approach	Purpose of approach: Mitigation of harmful effects of event	Purpose of approach: Avoidance of harmful effects of event
Outward	**Damage-sharing approaches**	**Joint problem-solving approaches**
Inward	**Isolating approaches**	**Evading approaches**

Figure 12.2 Categorisation of approaches to responding to desynchronisation.

the traditional logic of sequential task completion poses significant difficulties. The author's interviews with project professionals revealed the following mechanisms:

- Increased monitoring of another actor
- Increased quality-assurance efforts
- Use of subtle pressure to convince another actor
- Use of coercive (e.g. contract-based) power to convince another actor
- Rescheduling of activities
- Reallocation of resources/allocation of additional resources
- Imposing monetary penalties for being late
- Replacing another involved actor (termination of contract)
- Reallocation of responsibilities between involved actors
- Agreeing upon joint operating practices with another actor
- Using own resources to support another actor's task progress (helping another actor)
- Changing the project scope
- Assigning own resources to less critical tasks (slowing down own work)

The approaches as well as the patterns of using multiple approaches to complement each other are diverse, most likely limited only by creativity of individuals working with construction projects.

12.4 Performance Implications of Desynchronisation

Maintaining the synchronisation of the supply of materials has been linked to the success of manufacturing firms (Das and Goyal 1989). In a supply chain context, Cao and Zhang (2011) identified a relation between inter-organisational synchronisation and production output. In the context of project-based organising, Brusoni and Prencipe (2001) described 'achieving technological and organisational synchronisation' as a key role of the systems integrator firm, which is responsible for the delivery as a whole, and Dvir and Lechler (2004) demonstrated evidence of a relation between project changes and project success. Accordingly, the dynamic and inter-organisational process of organising how task interdependencies are managed and desynchronisation is responded to plays a highly central role in determining the performance of construction projects. In the context of construction, organising occurs at both the level of individual organisations (Pitsis et al. 2003) and across organisational boundaries (Sydow and Braun 2018). As this is the case, performance implications of desynchronisation need to be discussed separately from the perspective of individual firms and the project as a whole.

From the perspective of an individual firm, most of the implications of desynchronisation are unfavourable. Tasks often need to be rescheduled to accommodate interdependent tasks running behind schedule. Rescheduling always consumes resources and may be highly problematic, particularly if the affected tasks are situated on the critical path of the project. Resources used in projects are also typically time-delimited and may not be available – at least not without an increase in price – to cater to the revised schedule. In construction projects, the reallocation of resources is particularly challenging under periods of high demand, when resources are scarce. An organisation falling behind schedule may also suffer reputational damage, which could diminish its opportunities to be selected for forthcoming projects. Further financial damage may occur due to penalties for missing deadlines specified in contractual agreements. Importantly, not all implications of desynchronisation are unfavourable from the perspective of an individual firm. If it becomes evident that a specific firm is falling behind and cannot meet its obligations, the client may need to redistribute the work originally allocated to this firm. The client may, for example, establish subcontracting relationships with additional firms to secure the resources needed to get the project back on track. In these kinds of situations, the time pressure for securing resources is often considerable, and some of the subcontractors may be able to take advantage of this situation by increasing their prices. Moreover, a firm may benefit from desynchronisation when it lacks resources in the short term but can secure them in a cost-efficient manner at a later time. A firm struggling to meet its deadlines may be relieved to discover that other actors are falling behind and taking the blame for slowing down the project.

For a construction project as a whole, the implications of desynchronisation are predominantly unfavourable. Although an individual firm that is falling behind schedule may occasionally be able to catch up at its own cost, delays in critical tasks frequently cascade throughout the project organisation. A project task situated on the critical path can delay the entire project regardless of its scope. Some factors are external to project operations. As an indicator of this, during the COVID-19 pandemic, the global automotive industry suffered from a shortage of critical microchips, most of them less than €100 in value, resulting in temporary shutdowns of entire factories and the downgrading of production estimates for 2021 by hundreds of thousands of vehicles. To take another example, the United Kingdom experienced a reduced availability of lorry drivers following Brexit. This shortage compromised the availability of critical materials on construction sites throughout the country. Even when synchronisation can be restored without stopping the entire project, the costs of rescheduling tasks and reallocating the resources of multiple organisations unfavourably affect the performance of the project organisation.

Furthermore, instances of desynchronisation and who is responsible for it (and recovery) may foster an atmosphere of internal blaming and shaming, which could further hinder the flow of information and productivity. Although the desynchronisation of activities can often be remedied when sufficient (financial) resources are available, a more attractive option may be to scale down the project's scope. For example, ventilation equipment that cannot be acquired on time may be replaced with equipment with reduced performance specifications, or time may be cut from surface finishing, negatively influencing the aesthetic properties of the constructed building. Another problem related to desynchronisation is that it increases time pressure, potentially leading to suboptimal decisions and even mistakes in the construction process. Although increased time pressure may facilitate inter-organisational problem-solving and creativity, which could lead to innovations in working practices and technical details, it can be concluded that desynchronisation is very harmful for both productivity and the outcomes of construction projects. Table 12.2 summarises the performance implications of desynchronisation.

Table 12.2 Performance implications of desynchronisation.

	Potential negative implications of desynchronisation	**Potential positive implications of desynchronisation**
For individual firms	• Reduced productivity o Need for rescheduling of activities o Need for reallocation of resources • Reputational damage • Contractual penalty clauses • Increased time pressure	• Reduced time pressure • Opportunities for highly paid extra work/change orders
For the construction project	• Reduced productivity • Compromised teamwork culture (culture of blaming) • Increased pressure to reduce project scope • Higher probability of quality problems	• Potential for innovative solutions resulting from joint problem-solving activities

12.5 Factors Associated with Synchronisation in Construction Projects

Multiple factors at the organisational, project, and institutional levels influence a construction project's vulnerability to desynchronisation events. Drawing from the development of the Polaris missile system, Sapolsky (1972) introduced the use of parallel tasks, fallback strategies, and decentralisation as practices that can be used to reduce the risk of desynchronisation. Parallel tasks have been used in military development projects, such as the Manhattan Project (Lenfle and Loch 2010). This method reduces the project's susceptibility to losing synchronisation, because a failing task can simply be abandoned and replaced by a parallel task that is proceeding as planned; however, in the low-margin construction sector, the use of parallel tasks is rare due to its cost implications. Fallback strategies rely on the option of reducing the outcome's scope (for example, by removing planned features) if it becomes evident during the execution of the work that the original scope cannot be fully met by the current organisation. Finally, the decentralisation of authority increases the number of independent channels of communication available to the organisation, enabling actors to work out problems quickly and autonomously instead of bringing them to the attention of a centralised decision-making body. The decentralisation of control can be achieved through a number of practices. For example, Davies et al. (2009) discussed the widespread use of cross-functional teams in the construction of the Heathrow T5 terminal. Concentrating on intra-organisational new-product-development projects, Hoegl and Weinkauf (2005) verified the importance of two additional coordination mechanisms: project structuring and support and team interface management. The former mechanism involves the development of product integrity by achieving integration among various teams working on different modules in the project, whereas the latter mechanism involves inter-team communication and integration prior to freezing the project design.

The governance structure used in any construction project plays a central role in maintaining synchronisation, because it aims to align the interests and practices of organisational actors participating in the project (Ahola et al. 2014). This is accomplished by formulating shared rules and principles that are to be followed by the project actors. Some of these are included in project contracts, and some emerge during the course of the project as part of the continuous organising and self-organising that takes place amongst participating actors. As an example of a contractually defined rule, parties may agree on mechanisms for sharing bonuses paid to the project team or penalties payable by the project team. Project contracts also define the responsibilities of the participating organisations, which can range from very sharply defined to highly overlapping. If the responsibilities for project outcomes are at least partially shared across organisational boundaries, maintaining synchronisation becomes an issue of joint concern for the contracting parties. As an example of a principle that emerged during project execution, Ahola et al. (2017) discuss how the main contractor began to offer its subcontractors training related to welding highly specialised metal alloys to support them in maintaining the project schedule.

The roles of intangible and less formal elements, such as the project culture, are also significant. If the culture encourages actors to engage in filing claims and other types of zero-sum games that characterise construction in many parts of the world, motivating organisations to maintain synchronisation may be a very difficult problem. Instead, some actors may purposefully seek to desynchronise project activities for their gain at the cost of the productivity of the project as a whole. Generally, organisations that have worked together in the past and have established routines for collaboration are less likely to resort to opportunistic behaviour than are actors with no shared history (Eccles 1981; Sydow and Staber 2002).

Individuals occupying central roles in the project's organisation are highly important for maintaining synchronisation as well. Generally, persons working for the construction project should have a positive view of teamwork and mutual gains. Key individuals working on the project can be co-located to further enhance collaboration across organisational boundaries (Walker et al. 2017). It is vital that key persons possess sufficient decision-making power to negotiate agreements with other organisations included in the construction project – and that they do so in a flexible manner, as the need arises and without seeking approval from their superiors for every proposal they make. Advocates of a strict, hierarchical chain of command and treating subcontractors as mere servants are likely to have a negative influence on the motivation of the inter-organisational project team to maintain synchronisation.

The project-management tools used in the construction project also influence the actor's ability to maintain the synchronisation of activities. The tools used should allow real-time task-completion data and communication to flow across the entire network of organisations involved in the project. Preferably, programmes used for scheduling and progress monitoring should be harmonised throughout the project's organisation. The use of diverse and rich communication applications is likely to facilitate effective communication across organisational boundaries, although with the cost that an official record of all communication and inter-organisational agreements is likely to become impossible to create and maintain. Table 12.3 provides a summary of factors associated with the frequency of desynchronisation events in construction projects.

Table 12.3 Factors associated with the frequency of desynchronisation events in construction projects.

	Factors that decrease the frequency of desynchronisation events	**Factors that increase the frequency of desynchronisation events**
Organisational level	• Standardised processes for monitoring progress of tasks • Project managers with a high level of autonomy for decision-making (relative to their parent organisation) • Individuals with a personal preference for collaborating in key roles	• Ad hoc project-management practices that vary from person to person • Individuals with a personal preference for tight negotiations and zero-sum games
Project level	• Emphasis on expertise rather than hierarchy • Co-location of project's core team • Many participating firms with prior experience with collaboration • Project schedules shared across participating organisations • Use of cross-organisational teams • Main IT tools standardised across the project (e.g. scheduling, cost control, document storage) • Use of rich and flexible communication channels and applications • Open sharing of task-progress information across the entire supply chain • Gain- and pain-sharing agreements • Responsibilities that overlap organisational boundaries	• Large number of reciprocal task interdependencies • Emphasis on hierarchical chain of command • Widespread use of lump-sum contracts • Very long supply chains and low visibility • Communication restricted to approved channels only • Few participating firms with prior collaboration experience • Sharp and explicit delegation of the project's responsibilities
Institutional/ industry level	• Collaborative mindset dominant • Frequent use of alliance-based contracts • Disputes settled mostly outside courtrooms • Competition based mostly on superior deliverables	• Claims culture dominant • Late market cycle (economic activity at its peak, available resources scarce) • Disputes widely settled in courtrooms • Competition based mostly on price (very low margins)

12.6 Conclusion

In this chapter, I have described how the activities of organisations linked by interdependent activities can become desynchronised during a construction project and how synchronisation can be restored. I argue that although some individual firms may benefit from desynchronisation and may even seek to achieve it, its effects are in most cases highly unfavourable for individual firms and especially for the construction project as a whole.

Desynchronisation gives rise to disagreements and zero-sum games between firms participating in the construction project, wasting valuable project resources, risking the timely completion of the project and possibly undermining the functionality and quality of the project's deliverables. Because this is the case, devoting considerable effort to reducing the frequency of desynchronisation events at the level of both individual firms and the construction-project organisation as a whole makes good business sense.

Although many factors seem to relate to a construction project's vulnerability to desynchronisation events, two factors seem to play a particularly central role: attitudes and abilities of individual persons and the governance structure of the project. These factors are crucial as they influence the processes of organising and self-organising at both organisational and inter-organisational levels. Maintaining the synchronisation of tasks spanning organisational boundaries calls for open and ongoing communication and the willingness and ability to seek solutions that benefit the project as a whole. When problems arise, the joint intent of decision-makers must be finding the optimal solutions rather than identifying the parties responsible for the problem or the parties that will shoulder the costs of remedying it. In other words, key decision-makers must have a collaborative mindset and be granted enough leeway by their parent organisations to act accordingly. A collaborative orientation of decision-makers fosters the development of trust, which in turn facilitates the implementation of the project (Smyth and Edkins 2007). When a construction project is troubled, it is simply devastating for team morale to hear the project manager say the following words: 'I know this would be the right thing to do, but there is no way my supervisor would allow it'.

The governance structure of the project, which consists of both formal mechanisms, such as contractual agreements, and informal mechanisms, such as trust and informal gatherings of the project team, influences the processes of organising by establishing the rules that guide processes of organising across the entire inter-organisational project organisation. A key purpose of the governance structure is to ensure that the interests of the members of the inter-organisational project organisation are aligned and that all the members of the organisation benefit from the project's success or face harmful consequences if the project fails. As such, an important function of project governance is to recognise the need for continuous organising and create room for positive contributions, while trying to eliminate self-interested and other dysfunctional elements of organising. Regarding positive contributions, providing support for self-organising and mutual adjustment plays a particularly central role, as these processes are key for managing reciprocal task interdependencies crossing organisational boundaries. In practice, the project governance structure sets the limits in which these processes may operate. Formal governance mechanisms providing direction to purchasing practices in the project are also of importance. While arrangements for gain and pain sharing are likely to support task coordination, the use of lump-sum-based purchasing supports optimising at the level of a single organisation instead of the project as a whole. A governance structure tailored to the project can mitigate desynchronisation in two ways. First, it can reduce the frequency with which desynchronisation events occur. Second, when such events occur, it can ensure that all actors are motivated and able to engage in actions required to restore the project's synchronisation.

Previous studies of the delivery of complex projects (Brusoni and Prencipe 2001; Hobday et al. 2005) have highlighted the central role of the systems integrator firm – typically the main contractor in a construction project – in achieving technical as well as organisational integration. Although I do not question this, I argue that the synchronisation of activities across organisational boundaries cannot become the sole responsibility of a single organisation, even one that is highly resourceful and occupies a powerful position in the project organisation. Maintaining synchronisation must be viewed as the responsibility of each person in each organisation taking part in the construction project.

I conclude this chapter with a brief discussion of what makes for a world-class ice hockey team. First, the team requires highly skilled players. For example, the players need to have excellent skating skills, very high situational awareness, and the skill of passing the puck to a teammate without allowing the opposing team to intercept it. However, even if a team could recruit the best 20 players in the world, it would not necessarily be the strongest team in the world. To win, the team must work together seamlessly, flawlessly executing complex manoeuvres involving perfectly synchronised activities of multiple players with complementary roles and abilities. Although it is necessary to rehearse formal game plans and moves, there is no substitute for responsiveness to the context and run of play. Finally, to remain competitive over a longer period of time, the team must receive the support and resources needed to maintain the commitment of the players and the coach and to continuously attract new talent.

Analogously, a successful construction project comprises talented, motivated, and cooperation-oriented individuals working for firms driven by integrity, led by strong project management and possessing the necessary tools. These firms are joined via a governance structure that aligns the goals of the participating firms, rewards collaboration, and detects and punishes opportunistic behaviour and zero-sum games. Such a project has the capacity to operate like a well-oiled machine in which organisations work together in a synchronised and efficient manner.

References

Ahola, T., Ruuska, I., Artto, K., and Kujala, J. (2014). What is project governance and what are its origins? *International Journal of Project Management* 32 (8): 1321–1332.

Ahola, T., Vuori, M., and Viitamo, E. (2017). Sharing the burden of integration: an activity-based view to integrated solutions provisioning. *International Journal of Project Management* 35 (6): 1006–1021.

Assaf, S.A. and Al-Hejji, S. (2006). Causes of delay in large construction projects. *International Journal of Project Management* 24 (4): 349–357.

Brusoni, S. and Prencipe, A. (2001). Unpacking the black box of modularity: technologies, products and organizations. *Industrial and Corporate Change* 10: 179–205.

Cao, M. and Zhang, Q. (2011). Supply chain collaboration: impact on collaborative advantage and firm performance. *Journal of Operations Management* 29 (3): 163–180.

Chan, D.W. and Kumaraswamy, M.M. (1997). A comparative study of causes of time overruns in Hong Kong construction projects. *International Journal of Project Management* 15 (1): 55–63.

Das, C. and Goyal, S.K. (1989). A vendor's view of the JIT manufacturing system. *International Journal of Operations & Production Management* 98: 106–111.

Davies, A., Gann, D., and Douglas, T. (2009). Innovation in megaprojects: systems integration at London Heathrow Terminal 5. *California Management Review* 51 (2): 101–125.

Donaldson, L. (2001). *The Contingency Theory of Organizations*. Sage.

Dvir, D. and Lechler, T. (2004). Plans are nothing, changing plans is everything: the impact of changes on project success. *Research Policy* 33 (1): 1–15.

Dyer, J. (1997). Effective interfirm collaboration: how firms minimize transaction costs and maximize transaction value. *Strategic Management Journal* 18 (7): 535–556.

Eccles, R. (1981). The quasifirm in the construction industry. *Journal of Economic Behavior and Organization* 2 (4): 335–357.

Galbraith, J.R. (1973). *Designing Complex Organizations*. Addison-Wesley.

Gersick, C.J. (1988). Time and transition in work teams: toward a new model of group development. *Academy of Management Journal* 31 (1): 9–41.

Hällgren, M. and Maaninen-Olsson, E. (2005). Deviations, ambiguity and uncertainty in a project-intensive organization. *Project Management Journal* 36 (3): 17–26.

Hellgren, B. and Stjernberg, T. (1995). Design and implementation in major investments – a project network approach. *Scandinavian Journal of Management* 11 (4): 377–394.

Hobday, M., Davies, A., and Prencipe, A. (2005). Systems integration: a core capability of the modern corporation. *Industrial and Corporate Change* 14 (6): 1109–1143.

Hoegl, M. and Weinkauf, K. (2005). Managing task interdependencies in multi-team projects: a longitudinal study. *Journal of Management Studies* 42 (6): 1287–1308.

Lenfle, S. and Loch, C. (2010). Lost roots: how project management came to emphasize control over flexibility and novelty. *California Management Review* 53 (1): 32–55.

Lundin, R. and Söderholm, A. (1995). A theory of the temporary organization. *Scandinavian Journal of Management* 11 (4): 437–455.

Mandják, T. and Veres, Z. (1998). The DUC model and the stages of the project marketing process. In: *14th IMP annual conference proceedings*, vol. 3, 471–490. Turku School of Economics and Business Administration.

Pitsis, T.S., Clegg, S.R., Marosszeky, M., and Rura-Polley, T. (2003). Constructing the Olympic dream: a future perfect strategy of project management. *Organization Science* 14 (5): 574–590.

Sanchez, R. and Mahoney, T.J. (1996). Modularity, flexibility and knowledge management in product and organization design. *Strategic Management Journal* 17: 63–76.

Sapolsky, H.M. (1972). *The Polaris System Development. The Eisenhower Energy Policy: Reluctant InterventionReluctant Intervention*. Harvard University Press.

Smyth, H. and Edkins, A. (2007). Relationship management in the management of PFI/PPP projects in the UK. *International Journal of Project Management* 25 (3): 232–240.

Söderholm, A. (2008). Project management of unexpected events. *International Journal of Project Management* 26 (1): 80–86.

Sydow, J. and Braun, T. (2018). Projects as temporary organizations: an agenda for further theorizing the interorganizational dimension. *International Journal of Project Management* 36 (1): 4–11.

Sydow, J. and Staber, U. (2002). The institutional embeddedness of project networks: the case of content production in German television. *Regional Studies* 36 (3): 215–227.

Thompson, J.D. (1967). *Organizations in Action: Social Science Bases of Administrative Theory*. New York: McGraw Hill.

Tushman, M.L. and Nadler, D.A. (1978). Information processing as an integrating concept in organizational design. *Academy of Management Review* 3 (3): 613–624.

Uzzi, B. (1997). Social structure and competition in interfirm networks: the paradox of embeddedness. *Administrative Science Quarterly* 42 (1): 35–67.

Walker, A. (2007). *Project Management in Construction*, 5e. Blackwell Publishing.

Walker, D.H., Davis, P.R., and Stevenson, A. (2017). Coping with uncertainty and ambiguity through team collaboration in infrastructure projects. *International Journal of Project Management* 35 (2): 180–190.

Winch, G. (1989). The construction firm and the construction project: a transaction cost approach. *Construction Management and Economics* 7 (4): 331–345.

Woodward, J. (1965). *Industrial Organization: Theory and Practice*. Oxford University Press.

13

Organising Beyond the Hierarchy – A Network Management Perspective

Huda Almadhoob

Department of Architecture and Interior Design, University of Bahrain, Bahrain

13.1 Introduction

The combination of interdependence and uncertainty in large infrastructure projects provides a challenging environment for successful project delivery. This chapter builds on the work of Pryke (2017) in managing networks in project-based organisations, making the point that project delivery relies upon a range of activities that are essentially hidden from view, that are neither procured contractually nor codified in professional standards or governance arrangements. Hence, it emphasises the conceptualisation of construction project design and delivery systems as evolving self-organising networks, which form as a response to such environmental conditions (Pryke 2012; Almadhoob 2020).

In an attempt to extend our knowledge at the theory–practice interface of project management, the broader ontological debate adopted here is that projects are delivered through social networks, comprising of multiple actors responding to the pressures of finding and disseminating information in a highly uncertain environment (Pryke et al. 2018). These networks evolve and naturally decay over time, influenced by actors' interactions and their network position. This highlights the pervasiveness of change and the transformational nature of organisations and networks (Tsoukas and Chia 2002). From a project management perspective, this approach '*views managing social relationships as a means to manage and add value to, and through, projects*' (Smyth and Morris 2007, p. 425). Philosophically, this focuses on a 'becoming' ontology, which is primarily subjective and 'constructs' project reality based on an interpretivist epistemology (Pasian 2015; Magoon 1977; Tsoukas and Chia 2002; Langley and Tsoukas 2017). From this perspective, projects and their coordination processes are regarded as social phenomena that are continually being created from consequent actions of those actors involved (Bryman 2016; Grove et al. 2018). This view emphasises issues that reflect the ever-changing and evolving nature of the project environment (Pasian 2015) and the intrinsic flux of human action (Tsoukas and Chia 2002; Langley et al. 2013; Langley and Tsoukas 2017). Therefore, this perspective places huge importance on social interactions as the primary locus of social order rather than using

Construction Project Organising, First Edition. Edited by Simon Addyman and Hedley Smyth.

fixed hierarchical models (Blomquist et al. 2010). Tsoukas and Chia (2002, p. 570) advocate that '*change is ontologically prior to organization*', while '*organization is an outcome, a pattern, emerging from the reflective application of the very same rules in local contexts over time*'.

The key contribution of this chapter for construction project organising is to focus on cluster analysis in networks, being the true representative of project functions that carry out the day-to-day tasks and duties. It demonstrates, through a case study, that cluster analysis can offer a tool to both academics and practitioners to better understand the dynamics of project delivery by investigating cluster configurations over time, identifying the types of practices that different clusters are engaging in, and, most importantly, how to support or restrain certain behaviours through network-based interventions. The importance of supporting certain interventions and behaviours is dependent on effective decisions being made on the ground, and to this effect, Jager and Janssen (2003) decision-making heuristics model is also presented as part of the contribution to this chapter.

13.2 Social Network Analysis

This chapter investigates London's Bank Station Capacity Upgrade (BSCU) Project, with the help of social network analysis (SNA) as an analytical method to aid in understanding project actualities. The analysis focuses on SNA capability to capture invisible functionality that usually cannot be obtained using traditional scientific methods of project management. By doing so, the focus shifts beyond the descriptive formal–informal dualism in project delivery organising to identify and analyse project function-related clusters. This entails moving away from relatively abstract conceptualisations of project hierarchies and activities, to a much more finely grained approach that is based on analysing the dynamics of communication and information exchange networks, and how it relates to reducing project uncertainty. The network concepts outlined below are found to be instrumental in understanding the functioning of large project networks and are relevant to the discussion and analysis provided next.

First, actors are '*discrete individual, corporate, or collective social units*' (Wasserman and Faust 1994, p. 17). Graphically, these are represented by nodes. In this chapter, this term is given to individuals involved in the project information exchange networks rather than firms. Second, relations are '*the collection of ties of a specific kind among members of a group. . .The defining feature of a tie is that it establishes a linkage between a pair of actors*' (Wasserman and Faust 1994, p. 20). Graphically, these are represented by links/connections. Third, to detect community structure in networks, modularity analysis is used. It divides the graph into modules/groups/clusters, based on the concentration of ties and how a node can be easily grouped with other sets of nodes (Blondel et al. 2008).

Fourth, network density that is '*a concept that deals with the number of links incident with each node in a graph*' (Wasserman and Faust 1994, p. 101). It can be expressed as the total number of links present between nodes in a given network in relation to the maximum number of links theoretically possible for that network (Pryke 2012). This measure indicates network connectivity, that is, the degree of network integration or fragmentation (Goddard 2009). It also indicates the speed at which information diffuses within a network, degree of reciprocity, trust, and cooperation between actors, and hence whether they have high levels of social capital or constraints.

Fifth, preferential attachment is another important feature in the study of networks; it is likened to the 'rich-get-richer' rule. It provides explanations for growing inequalities in the process of network expansion/growth (Powell et al. 2005). That is, new nodes joining the network preferentially establish links with nodes that already have a large number of links (Heylighen 2011; Powell et al. 2005). This means that such a network has a high level of centrality and is vulnerable to targeted attacks on the nodes that function as hubs but robust to the random removal of links. This is highly risky in information exchange networks as the removal of hubs may deteriorate the network by splitting it into separate parts that no longer communicate with each other (Heylighen 2011).

Finally, brokerage/betweenness centrality reflects how often an actor lies on the path between other nodes in the network (Freeman 1979). Betweenness can be interpreted as a measure of potential control over information as it quantifies how much an actor acts as an intermediary to others (Freeman 1979; Borgatti 2006; Prell 2012). Actors with high levels of betweenness are usually regarded as 'brokers' or 'gatekeepers', as they act as critical bridges, reducing the distance between any two nodes (Freeman 1979; Borgatti 2006). They can help to hold the network together significantly by reducing transaction costs, but at the same time, they can control the flow of information (Freeman 1979; Borgatti 2006; Prell 2012) by retaining or biassing the information (Freeman 1979; Hossain and Wu 2009) and creating unnecessary iterative interactions (Pryke 2017). The latter may lead to an actor having a 'dysfunctional prominence' and may then lead to a disruptive functioning of the network (Pryke 2017). Cross et al. (2002) suggest that project teams may become over-reliant on these specialised, knowledgeable individuals. As a result, they become overloaded, leading to poor project performance.

The case study description is provided next followed by the empirical analysis and findings.

13.3 The Case Study

13.3.1 Background

London's BSCU project is a complex inter-organisational infrastructure project and was fourth in line of a programme of major station capacity projects for Transport for London (TfL).[1] The project was led and managed by London Underground Limited (LUL) as the client, a wholly owned subsidiary of TfL. The primary purpose of the BSCU project was to accommodate a significant increase in passenger demand and expected further growth at Bank underground railway station. To explain the complexity of the project in terms of task interdependency and surrounding interfaces, Bank Station is the fourth busiest interchange on the London Underground Network and is considered to be one of the world's most complicated subterranean railway stations. It is a key gateway into the City of London financial district, which is a densely populated and heavily congested area. It is located in a

1 Transport for London (TfL) is the statutory body accountable for all public transport within the City of London in the United Kingdom. Its remit covers all modes of transport – road, river, and rail – within Greater London.

designated 'conservation area'[2] and interfaces with over 60 properties, ranging from brand new commercial office developments to seventeenth-century churches and tunnelling within metres of the Bank of England. More than 250 engineers and staff were working at the BSCU project on a daily basis, while the station stayed open to customers throughout the work. Most of the construction work was taking place below ground to minimise construction impact on the historically significant site, which interfaces with 31 listed buildings.

Additionally, with the objective of procuring better value while delivering projects, LUL pioneered the development of a procurement process known as Innovative Contractor Engagement (ICE). ICE aims to '*control costs, speed up the works and reduce the impact on the travelling public. It also reinforces the aspirations to be an intelligent, innovative and efficient client that can build strong relationships with the supply chain whilst delivering value*' (TfL 2014, p. 5). However, by adopting the ICE procurement approach, the management of the BSCU project has been made even more complex. Arguably, the most important task was to create and enable a project team with the capability to support the uncertainties of the project as well as of the ICE process (Addyman 2019; TfL 2014).

13.3.2 BSCU Project Formal Organisational Structure

The project's formal organisational structure at the detailed design stage was hierarchical with a limited degree of mixed teams and responsibilities. This was predominantly informed by the traditional contractual structure. However, on top of the contractual

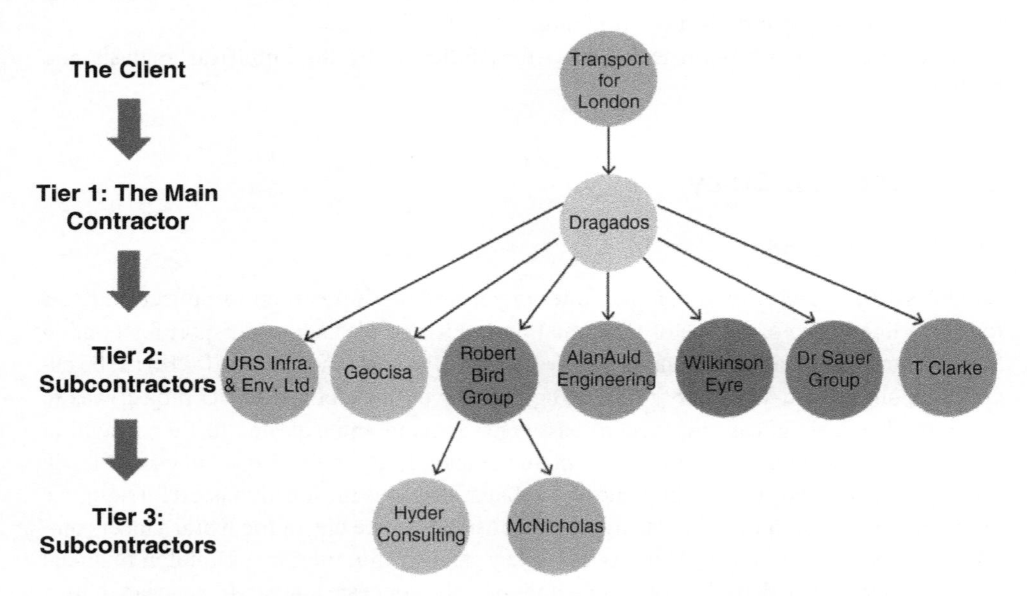

Figure 13.1 Contractual network between organisations involved in the BSCU project – 2015. Source: Pryke et al. (2017), Figure 3.1a, p.35 / With permission of John Wiley & Sons.

2 The term 'conservation area' applies to an area of special architectural or historic interest, the character of which is considered worthy of preservation or enhancement. It creates a precautionary approach to the loss or alteration of buildings.

arrangement, the project team layered a non-contractual collaboration agreement to help manage the interdependence, uncertainty, and complexity of the project (Addyman 2019). Overall, the contractual structure consisted of the client and three tiers of contractors, involving the contributions of more than 10 organisations and more than 250 personnel across various teams and roles. Figure 13.1 illustrates the three main tiers of the project supply chain. It places firms within the project contractual hierarchy, giving an indication of the lines of formal authority within the project organisation (Pryke 2012).

13.4 Empirical Analysis

Information exchange networks at two different stages of the detailed design of the project are investigated. The data represent networks of actors involved in the resolution of issues/events relating to the detailed design of the station box and the new ticket hall. Following a 'nominalist approach' (Pryke 2012), the project network boundary was defined. A summary of the main context surrounding the project at the two different stages of the detailed design is provided next. Around October 2014, while conducting the detailed design of the Whole Block Site,[3] possible pile clashes were identified. This issue was classified as a major risk as it could ultimately reduce the bearing capacity [4] of the piles, leading to potential cracking of the structural slabs and, therefore, serviceability and safety issues. This led to the need to develop an alternative design and construction sequences. Consequently, discussions with tier 2 subcontractors were held to determine possible solutions and mitigating actions. This was carried out over the rest of the year, resulting in some delays to key design dates. The general focus in the first stage, therefore, was redirected to provide sound structural solutions and mitigate the risks associated with progressing in the conceptual design. The context of the second stage was mainly concerned with the project control and management. By January 2015, the final decision on design changes, resulting from the identified pile clashes, was made. These changes were sent through change control by May 2015. In addition, around April 2015, new information on the sub-structure was received from London Underground, requiring the redesign and re-sequencing of the construction works to make time savings. Work in this period also focused on finalising proofs of evidence for the public inquiry and Transport and Works Act Order submission.

13.4.1 Clusters in BSCU Project Networks

Self-organising networks are characterised by a relatively short average path length and a high clustering, that is, the tendency of actors to concentrate their ties within certain groups (Baker 2014; Watts 1999). This stems from a distance-based cost structure and nodes that are closer or more similar (e.g. co-location, co-membership, co-participation, and sharing attributes) and find it cheaper to maintain links to each other leading to high clustering (Borgatti et al. 2014). These characteristics result in networks with unique properties of regional specialisation and efficiency (Watts 1999), hence providing the conditions to maintain the

3 Whole Block Site is located at Cannon Street London. It is the location of the entrance to Bank Tube Station on completion of the BSCU project.

4 The bearing capacity is the maximum load which a pile can carry without failure or excessive settlement of the ground.

network functionality to deliver projects (Borgatti et al. 2014). Therefore, the identification of these clusters is crucial in understanding the functionality of BSCU project networks.

Clusters were found by conducting modularity analysis, which was facilitated by Gephi software (Blondel et al. 2008). The analysis was conducted twice on the BSCU dataset, one for each stage. These were completed separately since the two stages are represented by two independent datasets. Seven self-organised clusters were detected at each stage (see Figures 13.2 and 13.3 for clusters found at stage 1 and stage 2, respectively). These clusters are inter-organisational and different to the organisational hierarchy identified earlier. Additionally, each cluster is composed of a number of highly specialised individuals, who are densely connected, and emerged with a distinct function in response to the need to gather and exchange information.

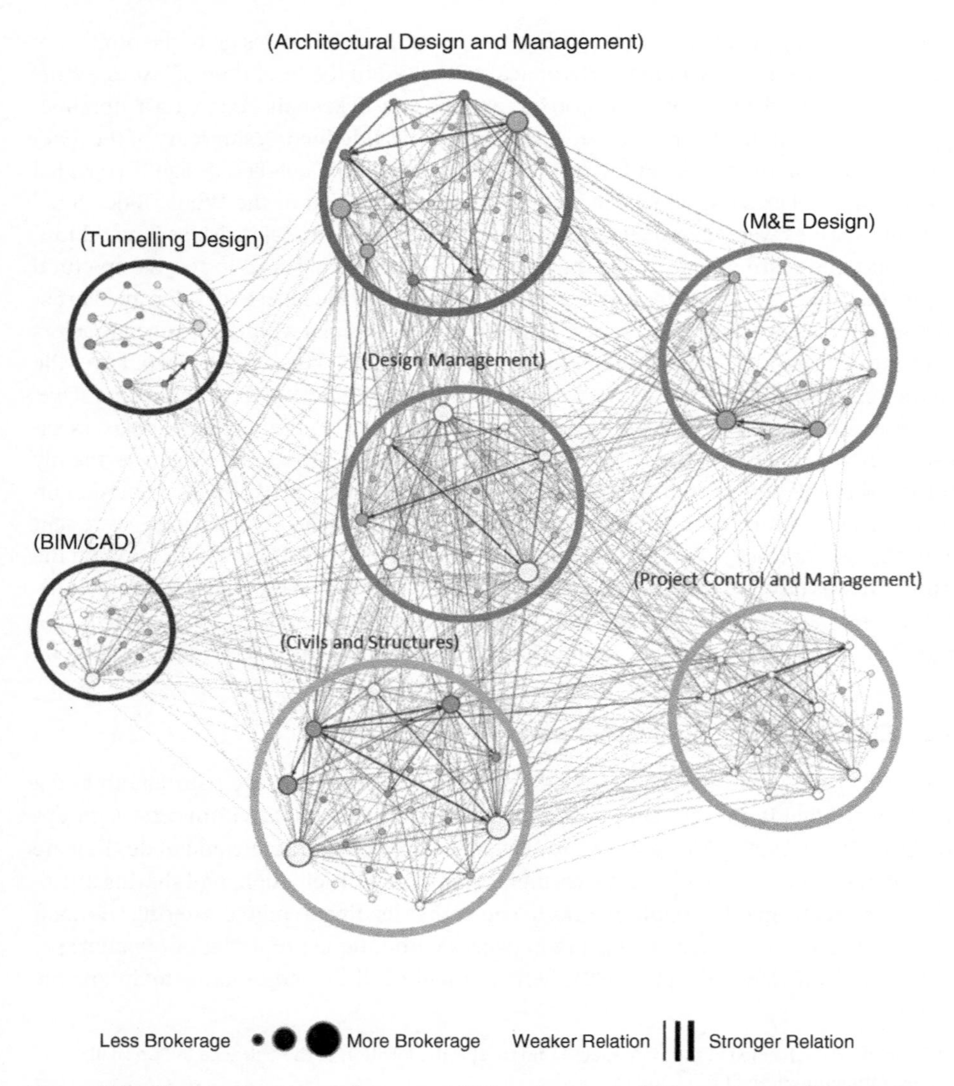

Figure 13.2 Project functional clusters at BSCU project stage 1. Nodes are sized by betweenness centrality. Ties are weighted by communication strength. *Source:* Adapted from Pryke et al. (2017).

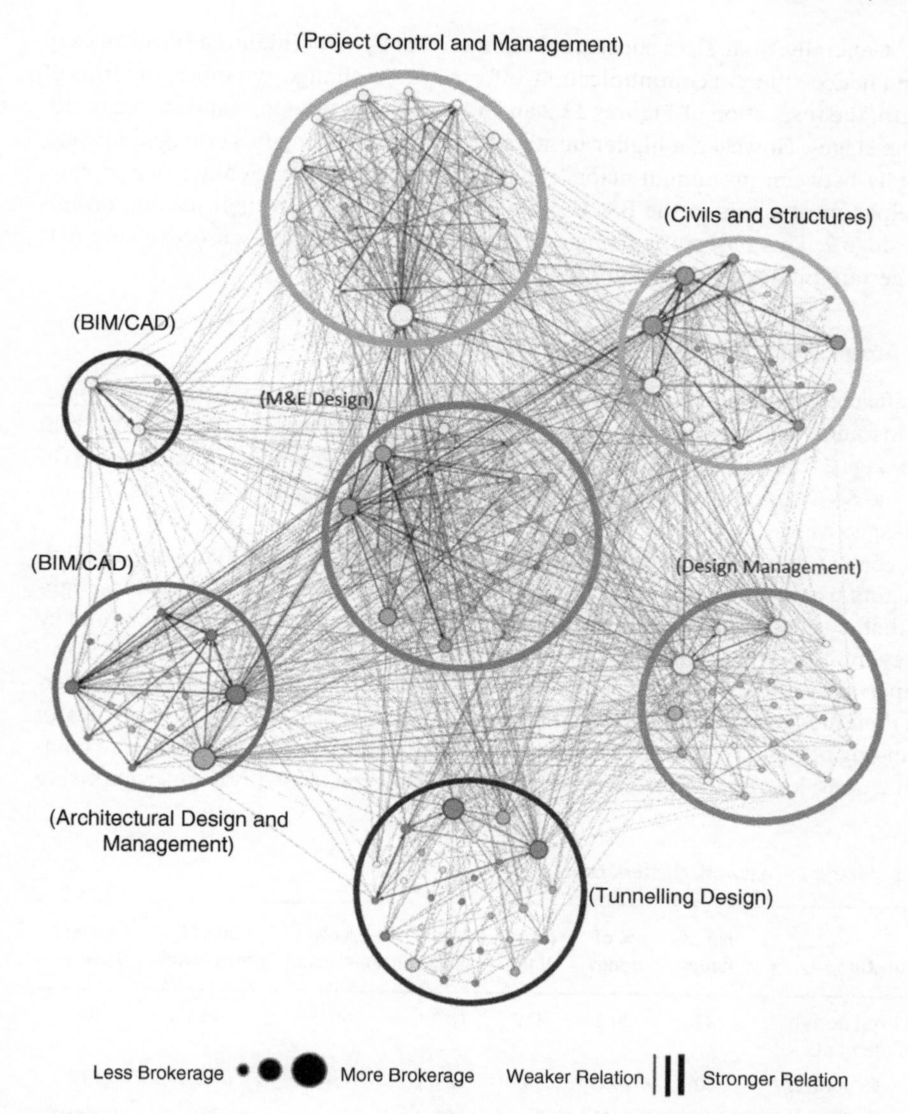

Figure 13.3 Project functional clusters at BSCU project stage 2. Nodes are sized by betweenness centrality. Ties are weighted by communication strength. *Source:* Adapted from Pryke et al. (2017).

While cluster detection is essentially an algorithm-based process, the classification of the clusters was based on their professional functional nature, namely: architectural design and management; civils and structures; project control and management; design management; Mechanical and Electrical (M&E) Design; building information modelling (BIM)/CAD; and tunnelling design. This was further confirmed through discussions with the BSCU project manager to ensure that identified clusters properly reflect how the project was delivered in relation to the contractual arrangement.

In Figures 13.2 and 13.3, nodes are sized by betweenness centrality. Ties represent the communication activity between the nodes and are weighted by communication strength.[5] Generally, node sizes suggest different levels of power (hierarchy) within each cluster and hence different communication/information exchange dynamics. In terms of tie strength, the inspection of Figures 13.2 and 13.3 reveals a predominance of weak ties at both the stages. However, a higher number of reciprocal strong ties emerged at stage 2, especially between prominent actors. This is explained by the 'rich-get-richer' rule, which helped most actors in the BSCU project networks maintain their leading prominence at stage 2. This observation suggests that even in distributed self-organising networks, a certain hierarchy or elite group will still be found.

13.4.2 Intra-Cluster Relationships and Decision-Making

The classification of project functions into clusters offers a tool to both academics and practitioners to understand the changes that occurred at each function as the project shifts from stage 1 to stage 2. Tables 13.1 and 13.2 provide a summary for the general characteristics (in terms of the SNA concepts described earlier) of each cluster at stage 1 and stage 2 of the project, respectively.

In this context, the composition of the clusters is investigated, identifying the distribution of actors based on their cluster affiliation. Table 13.3 shows this information, highlighting that clusters' compositions have changed between stage 1 and stage 2, mainly by way of new hires, leavers, and inter-transfers. The latter reflects the dynamics of clusters. More importantly, these inter-transfers facilitated cross-functional communication, leveraging on their previous relationships established at stage 1. Hence, the analysis provides an understanding of how clusters act as platforms to cope with project uncertainty and complex interactions, leading to reduced levels of conflict between too many actors comprising

Table 13.1 Stage 1 – network clusters' characteristics.

Cluster function	No. of nodes	% of nodes	No. of links	% of links	Links % intra-cluster	Links % inter-cluster	Cluster density
Architectural design and management	32	20%	410	18%	26%	74%	0.106
Civils and structures	30	19%	493	22%	31%	69%	0.177
Project control and management	26	16%	365	16%	37%	63%	0.209
Design management	26	16%	418	19%	25%	75%	0.158
M&E design	18	11%	307	14%	30%	70%	0.301
BIM/CAD	16	10%	126	6%	21%	79%	0.113
Tunnelling design	14	9%	112	5%	29%	71%	0.176

Source: Original.

5 The questionnaire data for communication frequency and quality scores were normalised and then multiplied. The outcome was used as a proxy for tie strength. This is in line with the method set out by Pryke (2012) and Pryke et al. (2017).

Table 13.2 Stage 2 – network clusters' characteristics.

Cluster function	No. of nodes	% of nodes	No. of links	% of links	Links % intra-cluster	Links % inter-cluster	Cluster density
Architectural design and management	18	9%	413	12%	14%	86%	0.186
Civils and structures	27	14%	593	17%	30%	70%	0.251
Project control and management	54	27%	666	19%	39%	61%	0.092
Design management	29	15%	530	15%	17%	83%	0.110
M&E design	37	19%	736	21%	33%	67%	0.181
BIM/CAD	4	2%	118	3%	5%	95%	0.500
Tunnelling design	28	14%	413	12%	27%	73%	0.149

Source: Original.

the entire project team (Gershenson 2007). In this sense, clusters manage the potential adversarial relationships between actors in a given network, who usually interact and make decisions locally before they reach consensus globally. This is particularly important to the understanding of network functions, dynamics, and the changing roles of actors within the BSCU complex project design network, as individuals collaborate and cultivate on informal relationships to deliver the project.

The analysis, therefore, highlights discrepancies between the roles procured contractually and the roles adopted in the self-organising clusters that evolved to discharge their functions. Such findings cannot be revealed by studying only the project contractual hierarchy, which often remains static, as represented in project governance documentation, making the identification of ongoing problems hard or too late if detected. Consequently, the effective project team configurations and any potential managerial interventions can be defined and replicated for managing other networks delivering projects. This is not to say that all the projects will have similar structures, but certain patterns and underlying processes behind the successful design and delivery of complex projects can be generalised (Pryke 2017). This is useful because many of the fundamental challenges in construction projects are the same whether the project in question is building a leisure centre, commercial office space, or rail infrastructure. Therefore, it is crucial to further study such clusters to look at the types of practices that different clusters are engaging in (e.g. leadership style, contractual arrangement, meeting timings, and forms of communication).

Due to space limitation, here, we look into the decision-making strategies employed in each cluster as the project, and thus, self-organising networks develop between the two stages. These are analysed by adopting the Jager and Janssen (2003) model, validation of which is confirmed by Holzhauer (2017), which emphasises the project's social processes. Fundamentally, the Jager and Janssen (2003) model defines uncertainty as an emergent state of social process rather than the view of uncertainty as a state of nature (time), under Winch (2002) model. Therefore, this means that uncertainty changes continually as a consequence of an individual or collective behaviour. The advantage of this perspective is that

Table 13.3 Distribution of nodes based on their cluster affiliation.

Cluster function		Where the actors come from based on their cluster affiliation at stage 1								
		Architectural design and management	**Civils and structures**	**Project control and management**	**Design management**	**M&E design**	**BIM/ CAD**	**Tunnelling design**	**New hires**[a]	**Total actors at stage 2**
Clusters at stage 2	Architectural design and management	12		1					5	**18**
	Civils and structures	1	20				2		4	**27**
	Project control and management	2	5	22	6		4		15	**54**
	Design management	10			14		2		3	**29**
	M&E design	3		1	5	18	1		9	**37**
	BIM/CAD						3		1	**4**
	Tunnelling design	1	2	1	1		2	14	7	**28**
Leavers[b]		3	3	1			2		N/A	**9** *(Total leavers)*
Total actors at stage 1		**32**	**30**	**26**	**26**	**18**	**16**	**14**	**44** *(Total new hires)*	**162/197** *(Total actors at stage 1/ Total actors at stage 2)*

a) These are the new actors joined BSCU project at stage 2.
b) Actors were part of stage 1, but they left the project at stage 2.Source: Original.

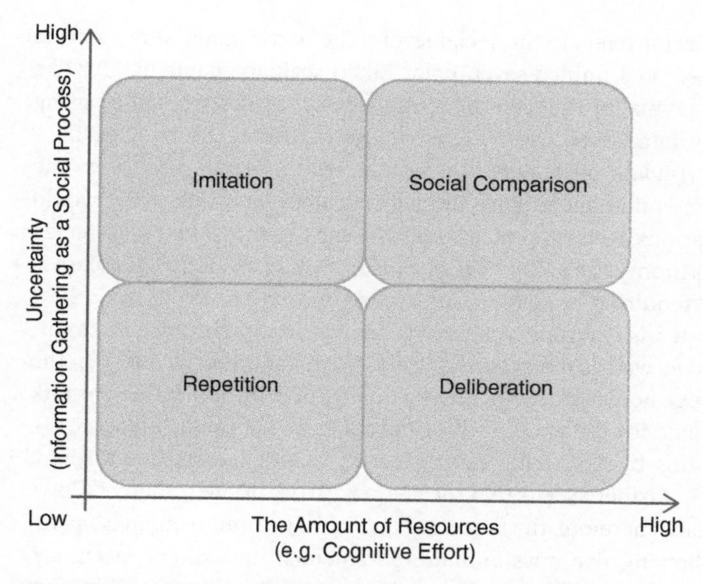

Figure 13.4 Decision-making heuristics organised in terms of uncertainty and amount of resources. *Source:* Adapted from Jager and Janssen (2003), Figure 3, p. 45.

it acknowledges the complex multi-level dynamics between the behavioural (local/individual) and structural (global/social) levels and, hence, the interdependence taking place within project teams (Lettieri et al. 2017; Madey et al. 2003).

This model organises various decision-making processes into four distinct heuristics,[6] based on the degree of uncertainty and amount of resources associated with making a decision, namely: imitation, social comparison, repetition, and deliberation. This model is illustrated in Figure 13.4.

The model is operationalised using project SNA data as follows:

The First Dimension – Uncertainty as a Social Process: The framework suggests that individuals will depend on their own experiences and abilities if they are certain and on others/social level if they are uncertain. Level of focus (local/global), therefore, will determine the degree to which social information is used in the decision-making process. The variable 'Frequency of Communication', collected as part of the study questionnaire, is used as the parameter to assess how often an actor reaches out to others to communicate and obtain information. The low frequency, therefore, means less communication activities, and hence, actors are satisfied with their local level of experience (i.e. low uncertainty requires a low degree of social information, and hence, an actor relies on his/her local level). On the other hand, the high frequency means high communication activities, that is, actors reaching out to others to seek for information (i.e. high uncertainty requires a high degree of social

6 Heuristics are powerful techniques in reducing complexity, defined as any approach to problem-solving that employs a practical method that is not guaranteed to be optimal, perfect, or rational, but instead sufficient for reaching an immediate goal (Myers 2010). Heuristics are shortcuts to speed up the process of finding a satisfactory solution; e.g. include using trial and error, a rule of thumb, and an intuitive judgement or common sense (Myers 2010).

information, and hence, an actor refers to the social level). The participants were asked to assign frequency scores based on a unified seven-point Likert scale to enable like-for-like comparison. Therefore, cut points for the uncertainty parameter are determined by using the mean value of frequency data. These are at 0.126 for stage 1 and at 0.133 for stage 2.

The Second Dimension – Amount of Resources Associated with Making a Decision: The basic idea of this dimension is that actors have limited resources, and thus, they would optimise their utility over various decision problems. This means an individual may decide a certain problem is not worth investing a lot of resources in, whereas another problem is of more importance, hence requiring more attention and resources. The decision on how much resources to be allocated is determined based on the nature, importance, and purpose of the problem in hand as well as the expected value (Jager and Janssen 2003). In the study of information exchange networks, the perceived quality of communication per link can be used as an approximate for the amount of resources to be allocated (Pryke 2017). Pryke (2017) has justified this by describing links between actors as conduits through which resources can flow. He further asserts that the characteristics of the resources/flows (e.g. amount) are quantifiable. Therefore, the perceived quality of communication is operationalised by using the following five questionnaire parameters: importance, accuracy, timeliness, clarity, and trust. The participants were asked to assign scores for these five parameters based on a unified seven-point Likert scale to enable like-for-like comparison. Hence, high quality of communication will need a lower amount of resources (e.g. a lower cognitive effort) for an individual to make a decision and vice versa. That is, the amount of resources associated with making a decision is calculated as the sum of these scores per communication link. The mean value of these data, thereafter, is used as the value for the cut points. These are at 0.488 for stage 1 and at 0.493 for stage 2.

The distribution of decision-making strategies across all clusters is illustrated in Figure 13.5. This involves defining the decision-making strategies employed by each actor and then grouping them based on their cluster affiliations. The results of each cluster are then expressed in relative terms.

Cluster analysis has yielded a number of important findings. First, the defined clusters are found to be inter-organisational and cross-functional, that is unrelated to the organisational/hierarchal chart. Therefore, these could be considered as an interpersonal relational capital that adds value to the BSCU project and organisations (Pryke et al. 2018; Kilduff and Krackhardt 2008). This is because they are neither formally/contractually procured by the client (TfL), nor are they owned by the organisations involved. This supports the argument of Pryke (2017) who finds that a highly interdependent environment is typically found in complex construction and engineering projects.

Second, BSCU project networks' clusters are characterised by having low-to-medium density values (see Tables 13.1 and 13.2). However, these values are higher compared to the density of whole network (stage 1 is at 0.055 and stage 2 is at 0.057). This suggests that, in complex projects, individuals tend to self-organise themselves into smaller groups/clusters to get the work done and minimise uncertainty.

Third, clusters in stage 2 reveal a different composition to that observed in stage 1. That is, some actors and ties were dissolved, and new or different ones emerged. This suggests that self-organising project networks are not static but dynamic. Within each function-based cluster, relationships evolve as project activities change, in reference to the level of

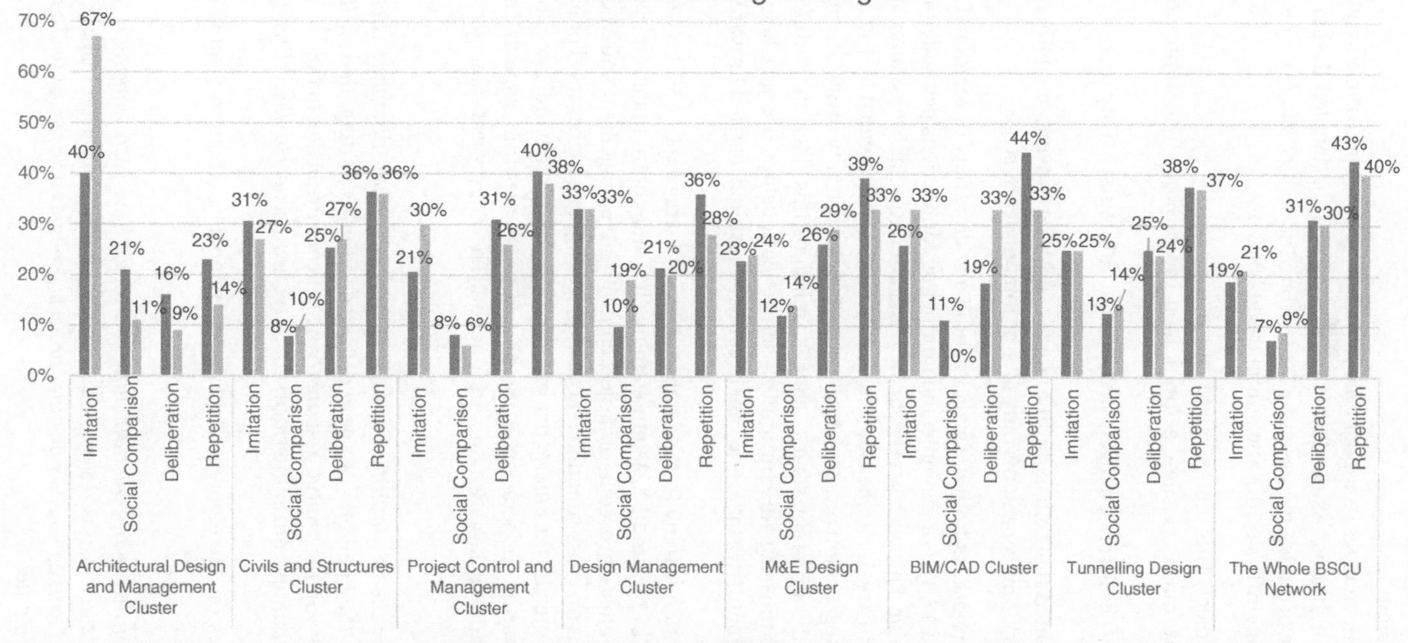

Figure 13.5 Distribution of decision-making strategies across all clusters at stage 1 and stage 2 of the BSCU project. The results of each cluster are expressed in relative terms, i.e. 'Total of each decision-making strategy in a given cluster'/'Total of all decision-making strategies for the same cluster.' *Source:* Original.

development of the design, and decay towards the end prior to completion or as a new activity comes onstream.

Fourth, the network changes its topography as necessary to adapt to the changing project requirements and conditions. As the BSCU project shifts from detailed design stage 1 to stage 2, the number of nodes increased by 22% (from 162 to 197 actors). Key structural changes are presented in Table 13.3 and summarised below:

- 9 actors left the project. This represents 6% of stage 1.
- 44 new hires were made (i.e. new actors joined the project). This represents 22% of stage 2.
- 50 inter-transfers (i.e. actors changed their cluster affiliation). This represents 31% of stage 2.
- 103 actors maintained their cluster affiliation (i.e. without moving to another cluster or leaving the project). This represents only 52% of total nodes at stage 2.

Fifth, informal networks at both the stages exhibit a wide variety of relational structures (see Figures 13.2 and 13.3). For example, the number of connections per node ranges from being dense to singular. This variety reflects the transient nature of project organisations in the construction industry. It can be said that this differential is created by the continuous negotiation and re-negotiation process of power (network centrality in SNA terms) to effect outcomes through the different stages. That is, relationships develop at different rates (hence the issue of time lags/discontinuity); some relationships are just being formed; some have reached maturity stage; and others could be residual, carried forward from previous project phases.

Sixth, communication activity (total number of links) increased by 55% between stage 1 and stage 2 (see Tables 13.1 and 13.2). Interestingly, the inter-communication was higher at stage 2 across all the clusters (except for project control and management and M&E design). This indicates a focus on cross-functional activities, leading to a higher level of integration. This is further supported by the emergence of new strong ties, which indicates an increased level of trust and mutual cooperation as the project progresses into stage 2. This finding suggests that integration is largely a human function. The TfL project manager reflects on this by asserting:

> *'When we saw the communities' data, they are quite multi-discipline clusters, apart from perhaps the M&E one. We were delighted with this finding, that's what you are trying to achieve in the project. You are trying to achieve function to function coordination* via *consultation team or interface people. We want people to speak directly together'.*

He further stresses on the element of path-dependence in creating a more integrated working environment:

> *'The main contractor and the tunnelling contractor have been working together since the tendering stage (late 2012–early 2013). They have a very good relationship and I think they had worked together on other projects probably overseas as well as in the UK, they knew each other'.*

The TfL project manager further highlighted that a number of interventions in the management of BSCU project were introduced at stage 2. Actor repositioning, for example the tier-1 contractor re-allocated resources to other designers to deal with the design delay and uncertainty regarding the buildability of proposed options, encouraged resource sharing, and arranged regular breakfast meetings to illustrate interventions. The management team found these to be particularly viable to enable more effective allocation of tasks, both gathering and dissemination of information, and improved decision-making and problem-solving.

Seventh, the ratio of inter-links to intra-links at both stages remained almost at the same level of 70:30% (see Tables 13.1 and 13.2). One may argue that the high percentage of inter-links means inadequate clustering since the actors are primarily engaged in an outward/cross-functional communication activity. However, this is not quite true because the algorithm used to produce these clusters is based on a density-based clustering method, which defines clusters as regions of higher density than the remainder of the dataset. The density, in turn, is determined by the ratio of the number of links an actor has to the total possible links an actor could have. The higher the ratio, the denser the set of connections. This explains the reason why intra-links are mainly formed by one-to-one relationships (that have fewer number of links but a high ratio of density; hence high clustering), whereas inter-links are mainly formed by one-to-many relationships (that have a large number of links but a low ratio of density, hence low clustering).

Finally, the distribution of decision-making strategies at the overall network level remained almost the same (see Figure 13.5). Repetition is the highest at 40%+, usually representing routines in projects which are necessary to provide consistency and stability and reduce mental effort (Becker 2004). Deliberation is at 30%, reflecting discussions taking place to exchange ideas and reach common grounds. Imitation and social comparison is around 20% and 10%, respectively, used to cope with higher levels of differentiation in the network (i.e. prominence and/or information asymmetry). This is common in highly complex construction projects, such as the BSCU project, as actors are highly specialised. That is to say, in complex projects, where the product of the project is essentially unique (having never been produced before), the mixed use of strategies reflects that actors are under pressure. Actors usually resort to routines in order to comply with contract conditions and project protocols, but at the same time the unique nature of complex projects demands adaptability in these routines. In terms of decision-making strategies, this conflict translates into an interplay between the desire of adopting autonomous actions (repetition) and/or cultivating social relationships.

In conclusion, for construction project organising, this chapter has illustrated the appropriateness of the network approach in studying the functioning of large infrastructure projects. Therefore, substantiating the need to move away from the focus on dyadic contractual relationships between firms towards a network management approach. SNA is of immediate relevance to project managers as it reveals dysfunctions in the project network and allows for the development of network-based interventions to support the development of collaborative relationships among network members. Such interventions can help managers facilitate successful project delivery through more effective decision-making strategies and, therefore, help reduce project uncertainty.

13.5 Conclusion

The case study project is characterised by a high level of uncertainty due to the involvement of an array of interfaces and interdependencies between its stakeholders who came from diverse disciplines. This chapter makes the point that project delivery relies upon a range of activities that are essentially hidden from view that is neither procured contractually nor codified in contract documentation. These activities are '*difficult to identify, difficult to quantify and therefore difficult to manage*' (Pryke 2017, p. 172). They essentially involve collaborative problem-solving with a high degree of self-organisation – a quality needed to overcome the fast-paced nature of infrastructure projects. It arises '*in response to individual actors' autonomy and motivation to seek and disseminate information, and in this way to discharge their contractually prescribed project roles* (Pryke et al. 2018, p. 37). The case clearly exhibits these features and confirms the nature of the 'hidden' relationships and dynamics, making them explicit in the process.

The analysis carried out in this chapter conceptualises the BSCU project as networks of relationships, which are mapped and interrogated utilising a number of SNA tools. In particular, it suggests identifying and analysing project function delivery clusters, moving away from a relatively abstract conceptualisation of project activities, and then suggesting an understanding of decision-making practices that these clusters are engaging in. Hence, providing a tool to managers to help support or restrain certain behaviours in projects.

The findings first suggest that as project actors must deal with uncertainty, they tend to form communication/information exchange clusters to ensure that project functions are carried out effectively. Typographically, clusters involve '*higher levels of density or connectivity than exists in the networks that surround these clusters*' (Pryke 2017, p. 147). This is of significance because such structure tends to indicate the development of various groups within the network, each with a different focus. The study of these clusters' topography, therefore, is important to the understanding of project networks and their management as it raises awareness to commence the setting of an agenda associated with the application of network theory. It enables project networks to be designed, replicated, and managed and perhaps sets project managers and clients thinking about interventions needed to achieve effective project execution – moving from the hidden towards overt understanding and organising.

Second, the investigation of BSCU networks identifies seven self-organising clusters in both the stages. It looks very specifically at interpersonal communications associated with design information exchange, classified based on project functions. By examining the topography of clusters and their intra-relationships, this chapter highlights the key changes at each cluster in terms of composition, prominence, communication activity, tie strength, and possible dysfunctions, suggesting that relationships evolve as project activities change.

Third, the findings show that clusters (and hence the overall network) are transient in nature; they are not static but dynamic, adapting to the changing project requirements and conditions. Findings further suggest that an element of path dependency affects the dynamics of self-organising networks and the creation and recreation of interaction patterns in construction projects. It also reveals a varying degree in terms of the presence of hierarchies within the network clusters. Hence, network structures are not about full elimination of hierarchies but rather the ability to facilitate the 'way things are done'. Therefore, power differentials (hence hierarchies) are necessary in order to activate the dynamics in the complex systems in response to higher levels of uncertainty/risks. That is to say, higher levels of decentralisation are not always a panacea as they can lead to some negative outcomes, for example, the hoarding of power/resources and not allowing them to be transferred to local levels. This was particularly found in the project control and management cluster, which exhibited resistance to the distribution of power outside of the contractual hierarchy. This means that organisational structures that are based on collaborative agreements can shape power relations significantly, yet do not determine them entirely. Contextual circumstances and individual behaviours are, therefore, crucial elements that should always be considered.

Fourth, decision-making strategies employed by individuals in BSCU networks were studied by adopting the Jager and Janssen (2003) model. This puts a great emphasis upon the extent to which social processes are adopted to reduce uncertainty. Actors in projects are generally attracted to routinised behaviour that offers consistency and stability, complying with contract conditions, and at lower cost. But, at the same time, the unique nature of complex projects demands adaptability in these routines, resulting in cultivating on social relationships.

To conclude, the following question is posed for future exploration: *how do these different self-organising project clusters fit together?* For construction project organising, exploring this question is essential to understand how higher coordination order emerges in networks from segregated and transient clusters.

References

Addyman, S. (2019). The Timing of Patterning or the Patterning of Timing? Organisational Routines in Temporary Organisations. PhD Thesis, University College London.

Almadhoob, H. (2020). At the interface: when social network analysis and supply chain management meet. In: *Successful Construction Supply Chain Management: Concepts and Case Studies* (ed. S. Pryke), 43–61. Oxford: Wiley.

Baker, W. (2014). Making pipes, using pipes: How tie initiation, reciprocity, positive emotions and reputation create new organizational social capital. In: *Contemporary Perspectives on Organizational Social Networks*, 57–72.

Becker, M.C. (2004). Organizational routines: a review of the literature. *Industrial and Corporate Change* 13 (4): 643–678.

Blomquist, T., Hällgren, M., Nilsson, A., and Söderholm, A. (2010). Project-as-practice: in search of project management research that matters. *Project Management Journal* 41 (1): 5–16.

Blondel, V.D., Guillaume, J.L., Lambiotte, R., and Lefebvre, E. (2008). Fast unfolding of communities in large networks. *Journal of Statistical Mechanics: Theory and Experiment* 10: P10008.

Borgatti, S.P. (2006). Identifying sets of key players in a social network. *Computational & Mathematical Organization Theory* 12 (1): 21–34.

Borgatti, S.P., Brass, D.J., and Halgin, D.S. (2014). *Social network research: Confusions, criticisms, and controversies*Contemporary perspectives on organizational social networks, 1–29. Emerald Group Publishing Limited.

Bryman, A. (2016). *Social Research Methods*, 5e. Oxford: Oxford University Press.
Cross, R., Borgatti, S.P., and Parker, A. (2002). Making invisible work visible: using social network analysis to support strategic collaboration. *California Management Review* 44 (2): 25–46.
Freeman, L.C. (1979). Centrality in social networks conceptual clarification. *Social Networks* 1: 215–239.
Gershenson, C. (2007). *Design and Control of Self-Organizing Systems*. Mexico: CopIt Arxives.
Goddard, S.E. (2009). Brokering change: networks and entrepreneurs in international politics. *International Theory* 1 (2): 249–281.
Grove, E., Dainty, A., Thomson, D., and Thorpe, T. (2018). Becoming collaborative: a study of intra-organisational relational dynamics. *Journal of Financial Management of Property and Construction* 23 (1): 6–23.
Heylighen, F. (2011). Rationality, complexity and self-organization. *Emergence: Complexity and Organization* 13 (1–2): 133–145.
Holzhauer, S. (2017). *Dynamic Social Networks in Agent-Based Modelling: Increasingly Detailed Approaches of Network Initialisation and Network Dynamics*. Kassel: Kassel University Press GmbH.
Hossain, L. and Wu, A. (2009). Communications network centrality correlates to organisational coordination. *International Journal of Project Management* 27 (8): 795–811.
Jager, W. and Janssen, M. (2003). The need for and development of behaviourally realistic agents. In: *Multi-Agent Simulation II* (proc. of the MABS 2002 Workshop, Bologna, Italy) (ed. J.M. Sichman, F. Bousquet and P. Davidson), 36–50. New York: Springer Verlag.
Kilduff, M., & Krackhardt, D. (2008). Interpersonal Networks in Organizations: Cognition, personality, Dynamics, and Culture (Vol. 30). Cambridge University Press.
Langley, A. and Tsoukas, H. (ed.) (2017). *The Sage Handbook of Process Organization Studies*. Thousand Oaks, CA: Sage.
Langley, A., Smallman, C., Tsoukas, H., and Van de Ven, A.H. (2013). Process studies of change in organization and management: unveiling temporality, activity, and flow. *Academy of Management Journal* 56 (1): 1–13.
Lettieri, N., Altamura, A., Malandrino, D., and Punzo, V. (2017). Agents Shaping Networks Shaping Agents: Integrating Social Network Analysis and Agent-Based Modeling in Computational Crime Research. In: *EPIA Conference on Artificial Intelligence*, 15–27. Cham: Springer.
Madey, G., Gao, Y., Freeh, V. et al. (2003). Agent-based modeling and simulation of collaborative social networks. In: *AMCIS 2003 Proceedings*, 237.
Magoon, A.J. (1977). Constructivist approaches in educational research. *Review of Educational Research* 47 (4): 651–693.
Myers, D.G. (2010). *Social Psychology*, 10e. USA: New York.
Pasian, M.B. (ed.) (2015). *Designs, Methods and Practices for Research of Project Management*. Gower Publishing, Ltd.
Powell, W.W., White, D.R., Koput, K.W., and Owen-Smith, J. (2005). Network dynamics and field evolution: the growth of interorganizational collaboration in the life sciences. *American Journal of Sociology* 110 (4): 1132–1205.
Prell, C. (2012). *Social Network Analysis: History, Theory and Methodology*. Sage.
Pryke, S. (2012). *Social Network Analysis in Construction. Supply Chain Management*, vol. 1. Oxford: Wiley.

Pryke, S. (2017). *Managing Networks in Project-Based Organisations*. Wiley.

Pryke, S., Badi, S., Soundararaj, B., & Addyman, S. (2017). Evaluation of collaborative teams for complex infrastructure projects through Social Network Analysis. [Online]. https://issuu.com/bartlettcpm/docs/ktp-report-2016 (Accessed on: 25th Sep. 2017).

Pryke, S., Badi, S., Almadhoob, H. et al. (2018). Self-organizing networks in complex infrastructure projects. *Project Management Journal* 49 (2): 18–41.

Smyth, H.J. and Morris, P.W. (2007). An epistemological evaluation of research into projects and their management: methodological issues. *International Journal of Project Management* 25 (4): 423–436.

Transport for London (2014). Innovative Contractor Engagement Report. [Online]. www.secbe.org.uk/content/panels/Report%20-%20Innovative%20Contractor%20Engagement%20Procurement%20Model%20-%20Bank%20Station%20Capacity%20Upgrade-6d5f2a.pdf [Accessed on: 04th July 2017].

Tsoukas, H. and Chia, R. (2002). On organizational becoming: rethinking organizational change. *Organization Science* 13 (5): 567–582.

Wasserman, S. and Faust, K. (1994). *Social Network Analysis: Methods and Applications*, vol. 8. Cambridge University Press.

Watts, D.J. (1999). Networks, dynamics, and the small-world phenomenon. *American Journal of Sociology* 105 (2): 493–527.

Winch, G. (2002). *Managing Construction Projects: An Information Processing Approach*. Wiley-Blackwell.

14

Procurement, Collaboration, and the Role of Dialogue

Simon Addyman

The Bartlett School of Sustainable Construction, University College London, UK

14.1 Introduction

The need for collaboration has become self-evident for the performance of the construction industry, seeking to positively influence the procurement and contractual arrangements through which construction projects are organised (Latham 1994; Egan 1998; Wolstenholme 2010). Stimulated by the client's choice of procurement model (Winch 2010; HM Gov 2020), construction clients and firms engage jointly in a variety of processes for creating construction project organisations, both transactionally and relationally (Bygballe et al. 2013). Over time, more than 15 000 varieties of procurement models have become available to clients showing that procurement is a dominant force in construction project organising, whatever the requirements and specification of the building (Hughes et al. 2006, p. 85).

In this chapter, I suggest that this large variety of models is a key factor that constrains more than it enables our performance as an industry. By this, I do not mean that there are too many models or that the industry is too fragmented, an often-cited problem. What I mean is that we tend towards an over-reliance on being able to *determine*, ex ante, the ideal fit between the procurement model and the intended outcome of the project. We spend less time fully appreciating the dynamic processes that participants engage in to accomplish those outcomes, within the boundaries of the chosen model, and arguably independently of the chosen procurement model.

Understanding these processes is necessary for knowing how to improve collaborative working practices for now and the future. This is because '*Collaboration as a process cannot be for its own sake, and if it is only a way to accomplish the present task, much of the potential is missed. Collaboration can create capacity for addressing not only the current problem but also those that follow. New ways of understanding collaboration can help us achieve these potentials*' (Feldman 2010, p. 159). For the future therefore, there is much to be gained from improving our understanding of the possibilities available to us when we engage effectively in collaborative working practices.

Construction Project Organising, First Edition. Edited by Simon Addyman and Hedley Smyth.

This chapter proceeds as follows: first, I define collaborative working practices as dialogic action. I show that from this perspective, organising is always an incomplete process, and hence, any procurement model cannot fully determine action patterns. I finish this section by presenting three aspects of incompleteness in construction project organising. Second, I discuss the role of language in organising and present Mikhail Bakhtin's Dialogism and his concept of chronotope as a lens for understanding dialogic action. Following this, I apply the chronotope as a method to analyse an empirical example of a dialogue from a design and build construction project in the UK rail industry. I then discuss these findings in relation to the incompleteness of construction project organising and what this means for collaborative working practices. I close the chapter by proposing an agenda for future theorising on the role of dialogue in informing our understanding of collaboration in construction project organising and how academia and practice may engage together on this.

14.2 Procurement and Collaborative Working Practices

In construction, there are a multiplicity of project organisational forms being created, often with the aim of improving performance. Some of the key drivers that influence these arrangements lie in the organisation of construction clients (Cherns and Bryant 1984), firms (Smyth 2018), supply chains (Pryke 2020), their governance arrangements (Winch 2001; Sydow and Braun 2018), institutional frameworks (HM Gov 2018), professional standards (APM 2019), digital technology (Chartered Institute of Building 2020), and, importantly for this chapter, the procurement and contractual arrangements (Hughes et al. 2006; HM Gov 2020).

During the 1990s in the UK, the government reports by Latham (1994) and Egan (1998) sought to redress the balance of the dominant 'bid low, claim high' culture by promoting a less adversarial and more collaborative approach (see Murray and Langford 2003 for a more detailed discussion of these reports). Reflecting on this move, a major study of procurement in the construction industry by Hughes et al. (2006) identified that supply chains are 'seriously engaging with collaborative working practices' (2006, p. x). Efforts continue to develop and implement more collaborative models in both academia (Mosey 2019) and practice (ICE 2017). Despite this, the traditional procurement route remains the dominant approach taken by UK clients (Oyegoke et al. 2009), and the capability of construction projects (and thus the industry more generally) to achieve effective collaboration remains in question (Farmer 2016; McKinsey 2017).

I conceptualise collaborative working practices as *dialogic actions* between participants and their artefacts within a shared social setting. I understand dialogic actions in construction project organising as being, through the use of language (spoken and written), the actions necessary for multiple interdependent participants to engage in a commercial transaction, communicate together, process information, and reduce uncertainty to accomplish their tasks, within a set of given boundaries (Addyman 2021; Higgins and Jessop 1965; Pryke 2017; Winch 2010). The meanings derived from these dialogic actions cannot be fixed and set out a priori in a procurement model. Following the work of Shotter (1995, 2008), I interpret the meanings derived from a model (in this chapter a procurement model) to be *transitory* and *anticipatory* in nature. That is, they enable participants to understand where they have come from, to where they are now (transitory), and how to capably *move on* into the future (anticipatory).

In this sense, a planned procurement model cannot fully determine the action patterns necessary to achieve project outcomes. This is because in an effort to enact these planned patterns, new meanings and patterns will emerge. This is an overt theoretical orientation towards the duality of patterning and performing in routine dynamics (Feldman and Pentland 2003; Feldman et al. 2021), which are essential for project organisations to become capable (Stinchcombe and Heimer 1985; Bygballe and Swärd 2019; Davies and Brady 2016; Davies et al. 2018).

To explore these dialogic actions, I draw on Mikhail Bakhtin's *Dialogism* (Holquist 2002) and his concept of *chronotope* (Bakhtin 1981), which I have previously applied to routine dynamics (Addyman 2019, 2021). From this perspective, the shared meanings derived from our dialogue are inherently social (constituted in a particular time and space), always incomplete, and always in the making (Holquist 1983; Bakhtin 1993; Shotter 2008, Tsoukas 2009). They are particularly incomplete in construction project organising because of the following:

1) By the very nature of the sequential processes involved in the initiation, design, and construction of a built asset, tender documentation put together as a part of the procurement model is necessarily incomplete (Pryke 2017).
2) The subsequent search for information to transition through the project life cycle is one of the generative mechanisms involved in the recreation of construction project routines (Addyman et al. 2020).
3) Firms and clients do not put in place organisational arrangements that help replicate interaction pan-project and on projects, for example through relationship management (Smyth and Edkins 2007).

This leads to the following research question for understanding procurement, collaboration and the role of dialogue in construction project organising: *How do our dialogic actions recreate shared meaning to achieve effective collaboration*? In the following section, I turn to the role of language, presenting Mikhail Bakhtin's Dialogism as an epistemology to explore these dialogic actions.

14.3 Language and Dialogue in Organisations

The use of language in organisations not only describes what we do but is used to create and recreate organisations and, in doing so, provides the possibility for the necessary coherence of change needed for organisations to function. Language provides an account of the relation between our intentions and our actions, where talk and action are seen as mutually constituted (Christensen et al. 2020; Bakhtin 2010; Searle 1969; Ford and Ford 1995). We use language (both spoken and written) to align our historical understanding of what we do through our professional roles, with the experiential nature of what we *actually* do in the here and now and how we orient towards future actions (Emirbyer and Mische 1998).

There are a variety of epistemological and methodological orientations to the study of language in organisations (Oswick et al. 2000; Boje et al. 2004; Grant et al. 2005). In this chapter, I follow a discursive approach, which focuses on action and the recreation of meaning over time (Tsoukas 2005, 2009). More specifically, I use Mikhail Bakhtin's Dialogism, '. . . a pragmatically oriented theory of knowledge; more particularly, it is one of several modern epistemologies that seek to grasp human behavior through the use humans make of language' (cited in Holquist 2002, p. 15). A number of organisational scholars working from a process perspective of organising have drawn on this theory because a dialogical approach enables us to explore how actors are relating to others and their social context of organising (Hatch and Ehrlich 2002; Cunliffe et al. 2004; Jabri 2004; Shotter 2008; Tsoukas 2009; Lorino and Mourey 2013; Cunliffe et al. 2014; Haley and Boje 2014).

For Bakhtin, 'language does not exist in the abstract, and is unregenerately active, and never settled or finalised. Bakhtin identified *utterance* as the basic unit of communication, bounded by speaking subjects in sociohistorically specific circumstances' (Holt 2003, p. 226, *emphasis added*). Utterances, words or bundles of words spoken or written in a given social setting, are the relation between speaker(s) and listener(s), not as a dyadic relationship, but triadic in the sense that the meaning of any utterance is constituted by the speaker, one or more listeners, and the social context within which the dialogue takes place (Shotter 1995). Importantly, the utterance by the speaker (or authors) is never original; it exists only in the ongoing stream of social life and, thus, acts as the boundary between what has come before and what might come after (Holt 2003; Holquist 1983). An utterance is an act of communication, 'which has the potential to mean, but contains no meaning in itself, its potential is realized through another's response' (Tsoukas 2009, p. 944). In this sense, the difference between our representation of what we plan to do, for example, a procurement model, and what we actually do by speaking or writing, is the place where shared meaning between participants is formed.

One particular concept Bakhtin used for analysing how this shared meaning comes together is what he termed the *chronotope,* literally translated as 'timespace' and originally used for understanding different genres of dialogue in literary theory (Bakhtin 1981; Bemong et al. 2010). The chronotope is a way of structuring together the different parts of a dialogue (a story, a conversation, or a text) between participants, at a particular time and in a particular place, into a relational whole that is sufficiently recognisable to those involved and enables them to capably *move on* from where they are now (Holquist 2002, p. 151). For example, when different project participants engage in a dialogue about a procurement model (speaking and writing), they connect together the different parts (for example strategy, roles, tender document, form of contract, works information, management plans, responsibility matrices, and digital platforms) in a way that enables an understanding of the their past, present, and future. When this understanding becomes shared, participants are able to accomplish their sequential interdependent tasks to achieve project outcomes (Shotter 2008). In this sense, they are creating the project capabilities to *move on*.

Each procurement model has a chronotope. Take, for example, the difference between a 'traditional' and a 'design and build' procurement model (Hughes et al. 2006). The chronotope of a traditional procurement model informs us of the expected separation between the designer and the constructor, with the client having two separate contracts with each party and where knowledge and information is exchanged from design to construction through formal artefacts, such as drawings and tender documentation. On the other hand, the chronotope of an integrated design and build model informs us of a single contract between a client and a design/construct provider, allowing for variety in the sequencing of activities and a multiplicity of forms of information exchange. Different participants will act differently in relation to these given chronotopes. Each chronotope will also have minor chronotopes (Addyman 2021, p. 198). For example, in the above two procurement models, contract administration will have its own discourse that has different parts that make a recognisable relational whole. In both of the above cases, a shared understanding is not given. The project specific time and context will require project actors to engage in a dialogue to create a shared understanding to be able to capably move on.

A number of organisational scholars have applied the chronotope (as concept and method) to investigate the dynamic processes of organising (Lorino 2010; Lorino and Tricard 2012; Musca et al. 2014). Lorino and Tricard (2012) used a construction project setting to identify eight categories of an organising chronotope. Through analysing the chronotope of a dialogue, we can identify the variety of possible actions that may fit with or deviate from the idealised action patterns intended from a planned procurement model, thus helping us understand the role of dialogic action in creating shared meaning. I built on the work of Lorino and Tricard (2012) and applied it to routine dynamics in my study of routine recreation in a construction project as it transitioned from design to construction under a design and build procurement model (Addyman et al. 2020). I further developed the chronotope as a method for exploring dialogic action in routine dynamics, and in Table 14.1, I provide a definition to the eight categories from the routine dynamics perspective (Addyman 2021).

Table 14.1 The generic categories of the routine chronotope.

Temporal frame	The inclusion of clock time (sequential and linear) and experienced time (past, present, future) involved in performing and patterning of the routine.
Spatial frame	The physical location(s) where action takes place and where actions' reference other spaces. This is both within one routine and across multiple routines.
Meaning-making principles	Understanding and anticipations of future conditions, taken from present or past routine patterns and performances, in achieving the goals of the tasks associated with the routine.
Roles and characters	The variety of participants and their role in the routine. Their role as formally prescribed by the organisation and its actual performance. Their role as interpreted by others.
Values	The personal and professional values that the participants express in their actions (doing and saying).
Crossing character	An action that contributes to the performance of more than one routine.
Artefacts	The artefacts that are associated with the routine.
Boundaries	The perceived and experienced boundaries of actions that constitute the repeatable and recognisable pattern of the routine.

Source: Addyman, 2021 / with permission of Cambridge University Press.

In the following section, I draw on this method to further highlight the dynamic and emergent nature of dialogue in a design and build construction project.

14.4 Applying Chronotope to the Analysis of Dialogue

I have selected the *contracting routine,* one of six routines that I identified in my original study on London Underground's Bank Station Capacity Upgrade Project (Addyman 2019). In this chapter, I have chosen the *contracting routine* as it fits with the discussion on procurement models. The goal of the contracting routine was to award the commencement notice for the construction stage of the design and build contract. It was a shared incentive target cost contract with a supporting non-contractual management protocol setting out how the parties aimed to collaborate. I analyse a brief excerpt of dialogue from the contracting routine, a minor chronotope. The setting is a project board meeting between senior executives of the client and contractor organisations, with the project directors from both parties in attendance. The dialogue is about enacting a standard contract clause on the application of indices for the measurement of inflation (indexation), which had been put into the contract at award stage. At that time, it represented an idealised structure for future actions to achieve an equitable share of risk from the impact of inflation. Yet, two years into the project, the applied calculations from regularly updated industry standards had started to pose future problems by eroding the contractor's profit margin. This concern had raised questions as to whether the contract terms should be amended to avoid any future potential conflict.

First, I present the dialogue in the form of a vignette. I analyse this in tabular format through the categories of the chronotope from Table 14.1 to give a picture of how the different parts of the dialogue create a recognisable relational whole. I then give a short narrative discussing how the dialogue and its shared meaning unfolds over time by highlighting the more dominant aspects in the categories of the chronotope from different participants' utterances (shown in italic in brackets). It is the understanding of these dominant aspects that helps create shared meaning (Figure 14.1).

While the exchange of information over the duration of the contract seeks to reduce uncertainty, in this instance, it has opened up greater uncertainty about the future (*time*) of the contractual risk share (*meaning making*) and alerts the participants (*roles*) to the location of the project in London (*space*).

The dialogue opens with the assertion by the client [CPD – L1 to L3] that the contract (*artefact*) will not be changed (*boundary*) and indicates antecedent (*time*) actions to consider such a proposition, which informs consequent actions, such as proposing a graphic (*artefact*) to help understand the possible future pathways. It places future obligations (*time*) on the Board (*roles*) to engage in a new dialogue with the data from this artefact as a new mediating tool (*meaning making, boundary*).

The dialogue proceeds with the client [CCD – L4 to L8] seeking to give further meaning to the decision on not changing the contract by opening this up as an opportunity for the contractor to engage with their supply chain (*roles/characters*) and take advantage of their international reach (*space*). This indicates there is no single direct consequent

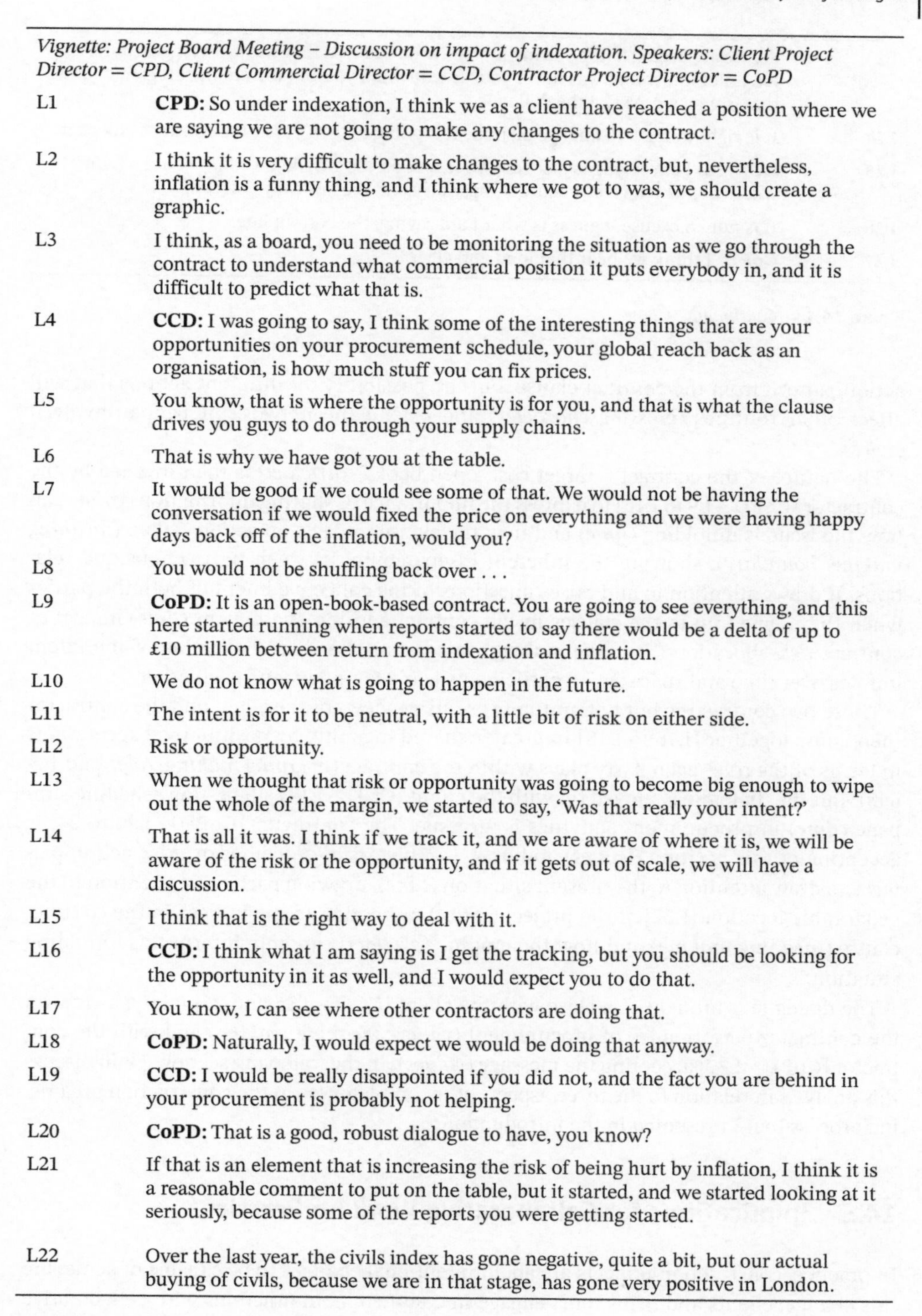

Vignette: Project Board Meeting – Discussion on impact of indexation. Speakers: Client Project Director = CPD, Client Commercial Director = CCD, Contractor Project Director = CoPD

L1	**CPD:** So under indexation, I think we as a client have reached a position where we are saying we are not going to make any changes to the contract.
L2	I think it is very difficult to make changes to the contract, but, nevertheless, inflation is a funny thing, and I think where we got to was, we should create a graphic.
L3	I think, as a board, you need to be monitoring the situation as we go through the contract to understand what commercial position it puts everybody in, and it is difficult to predict what that is.
L4	**CCD:** I was going to say, I think some of the interesting things that are your opportunities on your procurement schedule, your global reach back as an organisation, is how much stuff you can fix prices.
L5	You know, that is where the opportunity is for you, and that is what the clause drives you guys to do through your supply chains.
L6	That is why we have got you at the table.
L7	It would be good if we could see some of that. We would not be having the conversation if we would fixed the price on everything and we were having happy days back off of the inflation, would you?
L8	You would not be shuffling back over . . .
L9	**CoPD:** It is an open-book-based contract. You are going to see everything, and this here started running when reports started to say there would be a delta of up to £10 million between return from indexation and inflation.
L10	We do not know what is going to happen in the future.
L11	The intent is for it to be neutral, with a little bit of risk on either side.
L12	Risk or opportunity.
L13	When we thought that risk or opportunity was going to become big enough to wipe out the whole of the margin, we started to say, 'Was that really your intent?'
L14	That is all it was. I think if we track it, and we are aware of where it is, we will be aware of the risk or the opportunity, and if it gets out of scale, we will have a discussion.
L15	I think that is the right way to deal with it.
L16	**CCD:** I think what I am saying is I get the tracking, but you should be looking for the opportunity in it as well, and I would expect you to do that.
L17	You know, I can see where other contractors are doing that.
L18	**CoPD:** Naturally, I would expect we would be doing that anyway.
L19	**CCD:** I would be really disappointed if you did not, and the fact you are behind in your procurement is probably not helping.
L20	**CoPD:** That is a good, robust dialogue to have, you know?
L21	If that is an element that is increasing the risk of being hurt by inflation, I think it is a reasonable comment to put on the table, but it started, and we started looking at it seriously, because some of the reports you were getting started.
L22	Over the last year, the civils index has gone negative, quite a bit, but our actual buying of civils, because we are in that stage, has gone very positive in London.

Figure 14.1 Contracting routine vignette.

L23	So, we are seeing a huge swing in delta around civils that was worrying us quite a lot, and I know that was never your intent when you put the contract together, and that is why we started the dialogue.
L24	It is right to track it, and we all hope it is going to turn out as the contract expected.
L25	**CCD:** All I am saying is just keep your foot on the gas, doing what we are paying you to do, which is to procure things early and the best way you can.
L26	It is not an excuse, I guess is what I am saying, the bottom line.
L27	**CoPD:** I think we hear that loud and clear.

Figure 14.1 *(Continued)*

action pattern from the contract clause, but the possibility for different actions that will affect other routines (*crossing character*) and extend the network of people involved (*roles*).

The nature of the contract – target cost, open book – (*artefact*) is then invoked by the contractor [CoPD – L9 to L15] to express the importance of sharing information (*values*) on how the issue is unfolding (*time*) and how the clause impacts on actions taken (*meaning making, boundary*), showing the inherent inseparability between transactions and relations. It draws attention to and raises questions of the collective intent of both the parties when they signed up to the clauses in the contract and the consequent (*time*) impact of contract risk allocation (*meaning making*), suggesting the shared meaning of the intent unfolds over time and space as the participants come to understand each other.

These two contrasting but not mutually exclusive views of the client and the contractor then come together [L16 to L18] in greater shared meaning by positioning expectations in terms of the roles each party plays within the contract (*meaning making; roles/characters*). But the dialogue is moved on with the client [CCD – L19] suggesting schedule slippage (*time*) in procurement activities is an issue. The contractor [CoPD – L20 to 24] is accepting of this position (*values*) and draws on historical outcomes of prior action patterns to draw attention to the present situation (*time*), drawing particular attention to the geographic location [L22] of the project (*space*) in relation to the indices in the contract clause (*meaning making*) and how the graphic (*artefact*) can help inform this unfolding situation.

The dialogue is brought to a close with the client [CCD – L25 and 26] taking a view on the contractor performance of procurement (*values; meaning making*) and with the contractor [CoPD – L27] accepting the message (*values*). In the following section, I will discuss this analysis in relation to the three aspects of incompleteness in the construction organising process that I presented in the introduction.

14.5 Implications for Collaborative Working Practices

In practice, construction projects are about intentional change. To be capable of achieving this change, clients and firms must engage in commercial transactions with collaborative working practices that are able to recreate routines through the life of the project

(Mosey 2019; Hughes et al. 2006; Bresnen et al. 2005; Bygballe and Swärd 2019). A dialogic action perspective on these routines, through the lens of routine dynamics (Feldman et al. 2021), shows that the meaning of this intentional change is always incomplete and in the making – transitory and anticipatory – where change and stability are both an effortful and emergent accomplishment as participants position where they are in their process and how they might capably move on together (Pentland and Reuter 1994; Emirbyer and Mische 1998; Feldman 2000; Shotter 2008; Bygballe et al. 2021).

The empirical example here shows that the indexation contract clause in the design and build contract represented, at contract award, the planned actions for risk allocation from future price inflation. Yet, it could not determine the totality of actions to be taken, or indeed their meaning. Over time (two years), in an effort to enact the intended action patterns, new meanings emerged as the present situation of rising inflation in construction material and labour supply conflicted with the original intent of the clause (equitable share of risk). The chronotope as a concept and a method enabled the analysis of the dialogic actions between participants. It showed how utterances contained different configurations (drawing on dominant aspects in the chronotope categories) of their understanding to recreate their shared meaning and avoid misalignment. Looking back to the three aspects of incompleteness in the construction process signalled in the introduction, I present an example of each one.

First, after two years of working together, the dialogue points to some shared recent history of discussing the possibility of amending the clause, showing how any utterance is a moment embedded within the flow of a complex, multi-voiced, and open-ended dialogue (Holt 2003). The decision not to amend the clause had the possibility of constraining participants to continue to enact current action patterns. But in this example here, the dialogic action was a form of effortful relationship management as the senior management team applied the values attributed to their target cost contract and management protocol. Enabling possible pathways for different and potentially mutually beneficial patterns of action (Smyth and Edkins 2007). For example, the client points towards the timing of supply chain procurement and the connection to a wider set of organisations and people, while the contractor draws attention to the nature of the open book and the use of graphics to help share information. The project board played a key role in this relationship management.

Second, while the parties may have *agreed* the meaning of the clause at contract award, the data show that the team were still in a process of *agreeing* the meaning of that clause. This meaning has a correlation to the availability of information at a particular time and place, where specific actions related to the boundaries of search for information, influencing the multiplicity of possible future action patterns (Addyman et al. 2020). For example, there is the proposal for the production of an informative data graphic around which people can engage when discussing the potential misalignment between intended and potential outcomes. The actions involved in developing this graphic, and any subsequent actions taken from the information it provides, will create new dialogic actions that will help inform how far they have got in their understanding and how to capably move on. This search for and processing of information recreates organisational routines.

Finally, while the transaction between the client and the contractor has been set in the contract, the emerging need for further information and how that new information may be

Table 14.2 Indexation dialogue chronotope.

Temporal frame	Shared history of performing the contract gives the contracting parties new information, but this information opens a more uncertain future than the one planned for.
Spatial frame	The project/contract is enacted in London while the inflation indices are at a national level. The international contractor has global reach to influence its actions to manage the effects of the indices.
Meaning-making principles	The fixed contract clauses provide the axis for decision making and risk allocation but the emergence of the new 'graphic', showing impact of indices over time, is needed to aid decision making.
Roles	A multiplicity of new organisational and individual actors in the wider work context, beyond the dyadic client contractor relationship, are drawn into the actions of participants.
Values	An open book contract provides an opportunity for all parties to have visibility of information that will be shared, the clauses were fixed, but the interpretation was open and invited.
Crossing character	The action of 'no change' position to the contract clause on the indices contributed to other action patterns associated with indexation, such as patterns of action associated with procurement and scheduling of activities.
Artefacts	The fixed contract itself but also the development of a new artefact to help share knowledge and visualise the changing impact of the indices over time.
Boundaries	The incompleteness of information (what impact the indices are having) provides the boundaries to the dialogue (opening the decision to not change the clause) and judgements made by each speaker through the dialogue (closing with acceptance of potential future direction).

used extends both the number and the role of organisational participants, widening the network of people and organisations involved (Pryke 2017). For example, the contractor discusses how they are having a dialogue with their network of potential supply chain partners, trying to reduce the uncertainty by searching for new sources of information that may produce new opportunities. This search for more complete information extends the organisational network in a way not necessarily planned at contract award.

What we learn about collaboration in construction, when seen as dialogic action, is that the intended action patterns from the procurement model are but one set of possibilities for how the project outcomes may be achieved. These possibilities unfold as participants search for and exchange information through dialogue (both spoken and written). The chronotope method of analysis helps us see how these possibilities unfold as shared meaning emerges (Table 14.2). It shows how participants give emphasis to one or more aspects of chronotope categories to express their particular understanding in response to the boundaries of the procurement model. These different dialogic actions are continuously opening and closing opportunities and, thus, bracketing their action patterns over time (Lundin and Söderholm 1995).

Procurement models will always contain different aspects of incompleteness that require collaboration. Conventional approaches to construction project organising will need adapting if they are to positively influence collaborative working practices.

14.6 Conclusion

In construction project organising, the choice of procurement model remains a dominant force in how the project performs. Whichever procurement model is chosen from the myriad of options available (Hughes et al. 2006), construction project participants have the possibility of working collaboratively (Mosey 2019) in a way that is beneficial for both the present and future (Feldman 2010). To achieve this, they must engage effectively in a *dialogue*. The purpose of the dialogue is to create and recreate shared meaning through the patterning and performing of action (Tsoukas 2009; Caccatiori and Prencipe 2021; Bygballe et al. 2021). This is achieved by collectively understanding where they are now, where they have come from, and the possibilities of where they might be in the future (Emirbyer and Mische 1998; Shotter 2008). They do this by emphasising specific aspects important for their own interpretation of the information being shared within that social setting, with these aspects fitting to categories of the shared chronotope of the routine (Addyman 2021). This takes effort, with a need to recognise that new patterns of action will emerge if we are to achieve the intended outcome (Feldman et al. 2021). Two considerations emerge for academia and practice from this way of thinking.

First, construction project organisations need managers who can transition their organisational routines through the whole life cycle of the project (Addyman et al. 2020; IPA 2020). How they manage this routine (re)creation is a process of *speaking* (in our meetings, corridors, canteens, on site, and so on) and *writing* (for example, in emails, management plans, contracts/contract instructions, and monthly reports) in ways that enable and constrain the possibilities for effective action in a shared project context. They must allow time for participants to collectively interpret, from their shared and different perspectives, the information that is becoming available to them, or indeed its absence, and what this all means for accomplishing their tasks. For the project to become successful, this dialogic action must achieve the present task and capably move on towards the intended outcome, with minimal disruption. This needs effective relationship management through the formal structure and networks that connect the project–firm interface with the project.

Second, a dialogic action perspective of construction project organising also draws attention to how academia and industry might work together to develop new knowledge (Jones et al. 2021). A dialogical methodology requires the involvement of both practitioners and academics, with practitioners helping inform theoretical constructs and academics sharing in the organising of work as a collective multi-voice and multi-value inquiry (Lorino et al. 2011; Spence and du Gay 2021). This bridging of the gap between academia and practice is important because, when we talk about 'the industry', we must ensure that all voices and values in the dialogue are heard and play a role in developing the possibilities for improving collaborative working practices and the performance of the industry more generally (Shotter 2010).

With work and how we organise work becoming ever more uncertain and ambiguous as we point towards the middle of the twenty-first century (Barley and Kunda 2001), collaboration and procurement will continue to be influential forces in construction project organising. For collaboration to achieve its full potential for now and in the future, understanding the role of dialogic action in construction project organising can bring new and important knowledge for all stakeholders and the future performance of the industry.

References

Addyman, S. (2019). The Timing of Patterning or the Patterning of Timing? Organisational Routines in Temporary Organisations. PhD Thesis, UCL. http://discovery.ucl.ac.uk/10066709

Addyman, S. (2021). *Bakhtin's chronotope and routine dynamics*, chapter 14. In: *The Cambridge Handbook of Routine Dynamics* (ed. M. Feldman, B. Pentland, L. D'Adderio, et al.), 196–205. Cambridge: Cambridge University Press.

Addyman, S., Pryke, S., and Davies, A. (2020). Re-creating organizational routines to transition through the project life cycle: a case study of the reconstruction of London's Bank Underground Station. *Project Management Journal* 51 (5): 522–537.

APM (2019). *The Body of Knowledge*, 7e. UK: Association for Project Management.

Bakhtin, M.M. (1981). *The Dialogic Imagination: Four Essays* (ed. M.M. Bakhtin) (ed. C. Emerson and M. Holquist, trans.). Texas: University of Texas Press Slavic Series.

Bakhtin, M.M. (1993). *Toward a Philosophy of the Act*. University of Texas Press.

Bakhtin, M.M. (2010). *Speech Genres and Other Late Essays*. University of Texas Press.

Barley, S.R. and Kunda, G. (2001). Bringing work back in. *Organization Science* 12 (1): 76–95.

Bemong, N., Borghart, P., De Dobbeleer, M. et al. (2010). *Bakhtin's Theory of the Literary Chronotope: Reflections, Applications, Perspectives*, 1–227. Gent, Belgium: Academia Press.

Boje, D.M., Oswick, C., and Ford, J.D. (2004). Language and organization: the doing of discourse. *Academy of Management Review* 29 (4): 571–577.

Bresnen, M., Goussevskaia, A., and Swan, J. (2005). Organizational routines, situated learning and processes of change in project-based organizations. *Project Management Journal* 36 (3): 27–41.

Bygballe, L.E. and Swärd, A. (2019). Collaborative project delivery models and the role of routines in institutionalizing partnering. *Project Management Journal* 50 (2): 161–176.

Bygballe, L.E., Håkansson, H., and Jahre, M. (2013). A critical discussion of models for conceptualizing the economic logic of construction. *Construction Management and Economics* 31 (2): 104–118.

Bygballe, L.E., Swärd, A., and Vaagaasar, A.L. (2021). A routine dynamics lens on the stability-change dilemma in project-based organizations. *Project Management Journal* 52 (3): 278–286.

Cacciatori, E. and Prencipe, A. (2021). Project-based temporary organizing and routine dynamics. In: *Cambridge Handbook of Routines Dynamics* (ed. M. Feldman, B. Pentland, L. D'Adderio, et al.). Cambridge, UK: Cambridge University Press.

Chartered Institute of Building (2020). Reimagining construction: the vision for digital transformation, and a roadmap for how to get there. https://www.ciob.org/industry/policy-research/resources/digital-construction?gclid=EAIaIQobChMI_amK87Ky8QIVlLPtCh2XrQLmEAAYASAAEgI6GfD_BwE

Cherns, A.B. and Bryant, D.T. (1984). Studying the client's role in construction management. *Construction Management and Economics* 2 (2): 177–184.

Christensen, L.T., Morsing, M., and Thyssen, O. (2020). Talk–action dynamics: modalities of aspirational talk. *Organization Studies* 42 (3): 407–427.

Cunliffe, A.L., Luhman, J.T., and Boje, D.M. (2004). Narrative temporality: implications for organizational research. *Organization Studies* 25 (2): 261–286.

Cunliffe, A.L., Helin, J., and Luhman, J.T. (2014). *Mikhail Bakhtin (1895–1975).* In: *The Oxford Handbook of Process Philosophy and Organization Studies* (ed. J. Helin, T. Hernes, D. Hjorth and R. Holt). Oxford: Oxford University Press.

Davies, A. and Brady, T. (2016). Explicating the dynamics of project capabilities. *International Journal of Project Management* 34 (2): 314–327.

Davies, A., Frederiksen, L., Cacciatori, E., and Hartmann, A. (2018). The long and winding road: routine creation and replication in multi-site organizations. *Research Policy* 47 (8): 1403–1417.

Egan, J. (1998). *Rethinking Construction: Report of the Construction Task Force on the Scope for Improving the Quality and Efficiency of UK Construction.* London: Department of the Environment, Transport and the Regions.

Emirbayer, M. and Mische, A. (1998). What is agency? *American Journal of Sociology* 103 (4): 962–1023.

Farmer, M. (2016). *Modernise or Die: Time to Decide the industry's Future.* London, UK: Construction Leadership Council www.constructionleadershipcouncil.co.uk/wp-content/uploads/2016/10/Farmer-Review.pdf.

Feldman, M.S. (2000). Organizational routines as a source of continuous change. *Organization Science* 11 (6): 611–629.

Feldman, M.S. (2010). Managing the organization of the future. *Public Administration Review* 70: 159–163.

Feldman, M.S. and Pentland, B.T. (2003). Reconceptualizing organizational routines as a source of flexibility and change. *Administrative Science Quarterly* 48 (1): 94–118.

Feldman, M., Pentland, B., D'Adderio, L. et al. (ed.) (2021). *The Cambridge Handbook of Routine Dynamics.* Cambridge: Cambridge University Press.

Ford, J.D. and Ford, L.W. (1995). The role of conversations in producing intentional change in organizations. *Academy of Management Review* 20 (3): 541–570.

Grant, D., Michelson, G., Oswick, C., and Wailes, N. (2005). Guest editorial: discourse and organizational change. *Journal of Organizational Change Management* 18 (1): 6–15.

Haley, U. and Boje, D.M. (2014). Storytelling the internationalization of the multinational enterprise. *Journal of International Business Studies* 45 (9): 1115–1132.

Hatch, M.J. and Ehrlich, S. (2002). The dialogic organization, chapter 5. In: *The Transformative Power of Dialogue* (ed. N. Roberts). Emerald Group Publishing Limited.

Higgins, G. and Jessop, N.K. (1965). *Communications in the Building Industry.* London: Tavistock Publications.

HM Government (2018). Industrial strategy – construction sector Deal. https://assets.publishing.service.gov.uk/government/uploads/system/uploads/attachment_data/file/731871/construction-sector-deal-print-single.pdf

HM Government (2020). The Construction Playbook. https://assets. http://publishing.service.gov.uk/government/uploads/system/uploads/attachment_data/file/941536/The_Construction_Playbook.pdf

Holquist, M. (1983). Answering as authoring: Mikhail Bakhtin's trans-linguistics. *Critical Inquiry* 10 (2): 307–319.

Holquist, M. (2002). *Dialogism: Bakhtin and his World.* New York: Routledge.

Holt, R. (2003). Bakhtin's dimensions of language and the analysis of conversation. *Communication Quarterly* 51 (2): 225–245.

Hughes, W., Hillebrandt, P.M., Greenwood, D., and Kwawu, W. (2006). *Procurement in the Construction Industry: The Impact and Cost of Alternative Market and Supply Processes.* Oxon, UK: Taylor and Francis.

Infrastructure and Projects Authority, UK Government. (2020). Improving Infrastructure Delivery: Project Initiation Routemap Handbook. http://www.gov.uk/government/organisations/infrastructure-and-projects-authority

Institute of Civil Engineers (ICE), (2017). *Project 13 – From Transactions to Enterprises.* [Online]. Available from: www.ice.org.uk/knowledge-and-resources/best-practice/project-13-from-transactions-to-enterprises

Jabri, M. (2004). Change as shifting identities: a dialogic perspective. *Journal of Organizational Change Management* 17 (6): 566–577.

Jones, K., Mosca, L., Whyte, J. et al. (2021). The Role of Industry – University Collaboration in the Transformation of Construction, Transforming Construction Network Plus, Digest Series, No. 4

Latham, M. (1994). *Constructing the Team: Joint Review of Procurement and Contractual Arrangements in the UK Construction Industry.* UK: Department of the Environment.

Lorino, P. (2010). The Bakhtinian theory of chronotope (spatial-temporal frame) applied to the organizing process. In: *Proceedings of International Symposium on Process Organization Studies. Theme: Constructing Identity in and around Organizations*, 11–13.

Lorino, P. and Mourey, D. (2013). The experience of time in the inter-organizing inquiry: a present thickened by dialog and situations. *Scandinavian Journal of Management* 29 (1): 48–62.

Lorino, P. and Tricard, B. (2012). The Bakhtinian theory of chronotope (time-space frame) applied to the organizing process. Chapter 9. In: *Constructing Identity in and Around Organizations*, vol. 2 (ed. M. Schultz, S. Maguire, A. Langley and H. Tsoukas), 201–234. Oxford: Oxford University Press.

Lorino, P., Tricard, B., and Clot, Y. (2011). Research methods for non-representational approaches to organizational complexity: the dialogical mediated inquiry. *Organization Studies* 32 (6): 769–801.

Lundin, R.A. and Söderholm, A. (1995). A theory of the temporary organization. *Scandinavian Journal of Management* 11 (4): 437–455.

McKinsey Global Institute (2017). *Reinventing Construction: A route to higher productivity.* https://www.mckinsey.com/business-functions/operations/our-insights/reinventing-construction-through-a-productivity-revolution#

Mosey, D. (2019). *Collaborative Construction Procurement and Improved Value.* Hoboken, NJ, USA: Wiley-Blackwell.

Murray, M. and Langford, D. (ed.) (2003). *Construction Reports 1944–98.* Oxford, UK: Blackwell Science Ltd.

Musca, G., Rouleau, L., and Faure, B. (2014). Insights from the Crow's flight chronotope of the Darwin expedition. *Language and Communication at Work: Discourse, Narrativity, and Organizing* 4: 125.

Oswick, C., Keenoy, T.W., and Grant, D. (2000). Discourse, organizations and organizing: concepts, objects and subjects. *Human Relations* 53 (9): 1115–1123.

Oyegoke, A.S., Dickinson, M., Khalfan, M.M. et al. (2009). Construction project procurement routes: an in-depth critique. *International Journal of Managing Projects in Business* 2 (3): 338–354.

Pentland, B.T. and Rueter, H.H. (1994). Organizational routines as grammars of action. *Administrative Science Quarterly* 39 (3): 484–510.

Pryke, S.D. (2017). *Managing Networks in Project-Based Organisations*. Chichester: Wiley-Blackwell.

Pryke, S. (ed.) (2020). *Successful Construction Supply Chain Management: Concepts and Case Studies*. Wiley.

Searle, J.R. (1969). *Speech Acts: An Essay in the Philosophy of Language*, vol. 626. Cambridge University Press.

Shotter, J. (1995). In conversation: joint action, shared intentionality and ethics. *Theory & Psychology* 5 (1): 49–73.

Shotter, J. (2008). Dialogism and polyphony in organizing theorizing in organization studies: action guiding anticipations and the continuous creation of novelty. *Organization Studies* 29 (4): 501–524.

Shotter, J. (2010). Situated dialogic action research: disclosing "beginnings" for innovative change in organizations. *Organizational Research Methods* 13 (2): 268–285.

Smyth, H.J. (2018). Castles in the Air? The Evolution of British Main Contractors, www.ucl.ac.uk/bartlett/construction/news/2018/feb/castles-air-new-report-charts-management-issues-facing-british-owned-main-contractors.

Smyth, H. and Edkins, A. (2007). Relationship management in the management of PFI/PPP projects in the UK. *International Journal of Project Management* 25 (3): 232–240.

Spence, L.J. and du Gay, P. (2021). In praise of involvement. *Business & Society* 61 (4): 1–6.

Stinchcombe, A.L. and Heimer, C.A. (1985). *Organization Theory and Project Management: Administering Uncertainty in Norwegian Offshore Oil*. Oxford: Oxford University Press.

Sydow, J. and Braun, T. (2018). Projects as temporary organizations: an agenda for further theorizing the interorganizational dimension. *International Journal of Project Management* 36 (1): 4–11.

Tsoukas, H. (2005). Afterword: why language matters in the analysis of organizational change. *Journal of Organizational Change Management* 18 (1): 96–104.

Tsoukas, H. (2009). A dialogical approach to the creation of new knowledge in organizations. *Organization Science* 20 (6): 941–957.

Winch, G.M. (2001). Governing the project process: a conceptual framework. *Construction Management and Economics* 19 (8): 799–808.

Winch, G.M. (2010). *Managing Construction Projects*. Chichester: Wiley-Blackwell.

Wolstenholme, A. (2010). *Never Waste a Good Crisis: A Review of Progress since Rethinking Construction and Thoughts for our Future*. London, UK: Constructing Excellence https://constructingexcellence.org.uk/wolstenholme_report_oct_2009.

Author Biographies

Kirsi Aaltonen is an Associate Professor of Project Management at the University of Oulu, Oulu, Finland, where she heads the Project Business Research Team. Her current research interests include stakeholder management in complex projects, governance of inter-organisational projects, game-based learning methods, and institutional change in project-based industries.

Simon Addyman is an Associate Professor of Project Management at the Bartlett School of Sustainable Construction, University College London, London, UK, where he is Director of the MSc Project and Enterprise Management and Director of the Centre for Construction Project Organising. His primary research interests include understanding the nature of temporary organising and routine dynamics in construction, particularly how participants engage in a dialogue together to create value from projects.

Tuomas Ahola, DSc (Tech), is a Professor of Industrial Engineering and Management at Tampere University, Tampere, Finland. His research interests include management of complex inter-organisational projects, citizen participation, project governance, and sustainability in context of project deliveries.

Huda Almadhoob received her PhD degree in Construction Project Management from University College London, London, UK, in 2020. She is an Assistant Professor of Project-based Networks Management at the University of Bahrain, Zallaq, Bahrain. She is the Chairwoman of Architecture and Interior Design Department and the Director of Complex Networks Lab, which explores a range of theoretical and practical challenges through the lens of Networks organisation and management. Her primary research interests include project organising, organisational behaviours, and the application of social network analysis in the study of construction project-based networks.

Stewart Clegg is a Professor at the School of Project Management and the John Grill Institute at the University of Sydney, Camperdown, Australia, for Project Leadership and a Visiting Professor at the University of Stavanger Business School, Stavanger, Norway. He is recognised for his work in sociology, politics and power relations, organisation studies, and project management to all of which he has contributed substantially over his career.

Construction Project Organising, First Edition. Edited by Simon Addyman and Hedley Smyth.

D'Maris Coffman is Professor of Economics and Finance of the Built Environment and Head of Department at the Bartlett School of Sustainable Construction, University College London, London, UK. She is an Editor-in-Chief of *Structural Change and Economic Dynamics* and on the honorary editorial boards of *The Journal of Cleaner Production, Economia Politica*, and the *Chinese Journal of Population, Resources and Environment.* Distinguished Visiting Professor at Tsinghua University, Beijing China, a Visiting Professor at the University of Milan (Statale), Milan, Italy, and a Visiting Professor at the Renmin University of China, Beijing. She works at the interstices of economic geography, economic history, and infrastructure economics. Her main research interests include infrastructure economics, climate change economics, and energy transitions.

Andrew Davies has been the RM Phillips Freeman Chair and a Professor of Innovation Management at the Science Policy Research Unit (SPRU), University of Sussex Business School, since 2019. He started his career at SPRU, Sussex University, Brighton, UK, before moving to Amsterdam University, Imperial College London, and University College London (UCL). He returned to SPRU in 2019 and continues as an Honorary Professor at the Bartlett Faculty of the Built Environment, UCL. He has a long standing association as Visiting Professor at the Department of Business and Management, LUISS, Rome, Italy, and the BI Business School, Oslo. His current interests lie in innovation in complex projects and large scale infrastructure.

Meri Duryan is a Skills and Knowledge Operations Manager at Arup University, London, UK. At the time of writing, she was an Associate Professor of Enterprise Management at the Bartlett School of Sustainable Construction, University College London, London. In addition to her academic and publishing background, she is an APM-certified change management practitioner with over 20 years of expertise, delivering visionary change management and knowledge management strategies and solutions across High Tech, Education, Healthcare, Transportation, and Construction industries. Her research interests include systems thinking, organisational learning, service design, and occupational health, safety, and well-being.

Timothy Jaques, PMP, is a Serial Entrepreneur who lives in New York, USA. His latest venture, Teaming Worldwide, is creating innovative new platforms and business models in healthcare and the next generation of hospitals. He has co-authored two books including Achieving Project Management Success in the Federal Government, a 2010 PMI Book of the Year nominee. He is a contributor to numerous global standards including the International Project Management Association Individual Competence Baseline, version 4. He continues to explore and speak about the challenges of modern and complex projects.

Susanna Hedborg received her PhD degree in Business Administration from the KTH Royal Institute of Technology, Stockholm, Sweden, in 2022, with specialisation in project studies and urban development. She is currently a Postdoctoral Researcher at the Division of Industrialised and Sustainable Construction, Luleå Technical University, Luleå, Sweden. Her research interests include project organising and organisational routines in the built environment.

John Kelsey is an Associate Professor of Construction, Project Management, and Economics at the Bartlett School of Sustainable Construction at University College London. He has worked in India, Russia, Africa, and UK/France on the Channel Tunnel. His teaching and research areas have included construction planning, technology transfer, building information modelling, stakeholder engagement, rural housing, health and safety, and organisational culture.

Catherine Killen is a Professor specialising in project portfolio management (PPM) at the School of the Built Environment, University of Technology Sydney, Ultimo, Australia. Her influential PPM research ranges from practice-based studies exploring current approaches to experiments and trials of new methods. She is best known for her work on PPM as a dynamic capability that aligns projects with strategy and enables organisations to respond to changes in the environment.

Ole Jonny Klakegg received his PhD degree in Project Governance in 2010. He is a Professor of Project Management at the Norwegian University of Science and Technology, Trondheim, Norway. Throughout his whole career, he has alternated between practicing in the construction industry and working with research and teaching in the university. His current research interests include collaborative project delivery for enhanced value creation and sustainability.

Martin Loosemore is a Distinguished Professor of Construction Management at the University of Technology Sydney, Ultimo, Australia. He is an Honorary Professor at the University of Cape Town, Cape Town, South Africa, and a Fellow of the Chartered Institute of Building. He has acted as an Adviser and a Contributor to a number of Australia Federal Government Royal Commissions and State Government Inquiries into Building and Construction Industry reform. He is the founding partner of a successful social enterprise, which specialises in securing meaningful employment opportunities in the construction industry for people experiencing disadvantage. His research has been widely commercialised through a number of spin-out enterprises.

Bethan Morgan combines extensive management experience with substantial expertise in the teaching and research of management and digital transformation in businesses. She is involved in work relating to social and environmental value. She is holder of an EPSRC fellowship in the UK and was awarded her PhD degree in technological adoption and organisational change from University College London, London, UK. Her current interests include leadership for change, particularly in the construction industry.

Nils Olsson received his PhD degree in project flexibility from the Norwegian University of Science and Technology, Trondheim, Norway, where he is a Professor of Project Management. He has extensive experience as consultant, research scientist, and manager. His consulting experience includes Ernst & Young and Det Norske Veritas. His current research interests include emerging aspects of project management, ranging from governance to applications of artificial intelligence in projects contexts.

Eleni Papadonikolaki is an academic and consultant. Her research has been funded by Innovate UK, European Horizon programmes, and Project Management Institute. She is an Associate Professor at University College London, London, UK, and the Founding Programme Director of MSc Digital Engineering Management. Her research interests include intersection of management, social science, and engineering.

Shankar Sankaran is a Professor of Organizational Project Management at the School of the Built Environment, University of Technology Sydney, Ultimo, Australia. He is recognised for his work on project management, project leadership, project governance systems thinking, mega-projects, and action research. He has recently received the PMI 2022 Research Achievement Award with Ralf Müller and Nathalie Drouin.

Hedley Smyth is an Emeritus Professor at the Bartlett School of Sustainable Construction, University College London, London, UK. His primary teaching and research interests focus on the management of the project-based firm and main construction contractors in particular. This includes understanding the extent of management at the firm-project interface. Strategy and organisational behaviour provide two areas of research focus. Business development, health and safety, and well-being continue to be substantive areas of research. Project portfolio management is a further recent area that links back to strategy management and the firm-project interface.

Virpi Turkulainen works at the Haaga-Helia University of Applied Sciences, Helsinki, Finland. She received her D.Sc. in operations management at Aalto University. In the past, she has worked at the School of Business, University College Dublin, Dublin, Ireland, and the School of Science, Aalto University, Espoo, Finland. Her research interests include organisation design, organisational integration, management and organisation of project-based operations, and operations strategy.

Derek Walker is an Emeritus Professor in a continuing sessional position. He is a co-editor (with Professor Steve Rowlinson from Hong Kong University) of *The Routledge Handbook of Integrated Project Delivery* (2020) that is a 28-chapter book that he co-authored with a range of world experts on aspects of integrated project delivery. He has authored or co-authored over 70 book chapters, 11 books, and 120 refereed journal papers. He was a Founding Editor for the *International Journal of Managing Projects in Business* and an Editor for its first decade. His research interests include strategic project management collaboration, organisational learning, innovation diffusion, and project governance.

Alfons van Marrewijk is a Full Professor in Construction Cultures at the Faculty of Architecture and the Built Environment, Department of Management in the Built Environment, Delft University of Technology, Delft, The Netherlands. He is also an Adjunct Professor of Project Management at the BI Norwegian Business School, Oslo, Norway, and an Associate Professor at the Department Organization Sciences, Vrije Universiteit Amsterdam, Amsterdam, The Netherlands. In his academic work, he uses anthropological theories and methods for studying inter-organisational collaboration and cultural change in technically oriented organisations and complex mega-projects. He has published in numerous journals and works in close collaboration with public and industry partners in construction and infrastructure.

Jennifer Whyte is a Professor and the Head of the School of Project Management, Faculty of Engineering, University of Sydney, and the Director of the John Grill Institute for Project Leadership. She retains a position at Imperial College London, London, UK, where she was the Director at the Centre for Systems Engineering and Innovation and the Royal Academy of Engineering and a Laing O'Rourke Professor of Systems Integration at the Department of Civil and Environmental Engineering from 2015 to 2021. Her research interests include project leadership, innovation, and mega-projects, which has had significant impact on transforming industry and policy.

Wu Yanga is a last-year PhD student at the Bartlett School of Sustainable Construction, University College London, London, UK. She works as a teaching fellow at the School of Business, Operation and Strategy, University of Greenwich, London. Her primary research interests include organisational routine study in technology adoption process within the construction industry, project portfolio management and strategic alignment.

Jing Xu received her PhD degree in Construction and Project Management from University College London (UCL), London, UK. She is a Lecturer of Enterprise Management at the Bartlett School of Sustainable Construction, UCL. She is experienced in the management of project-based firms and their people. She participated and led research projects funded by Association for Project Management, Lloyds Register Foundation and UK Research and Innovation. Her recent research interests include care ethics in project management and its implications to project workers' well-being.

Index

Note: **Bold** page numbers refer to tables and *Italic* page numbers refer to figures.

Construction Project Organising, First Edition. Edited by Simon Addyman and Hedley Smyth.

d

q

r